AF556871

Tropical Mycology
Volume 1, Macromycetes

Tropical Mycology

Volume 1, Macromycetes

Edited by

Roy Watling,

Juliet C. Frankland,

A.M. Ainsworth,

Susan Isaac

and

Clare H. Robinson

CABI *Publishing*

CABI *Publishing* is a division of CAB *International*

CABI Publishing
CAB International
Wallingford
Oxon OX10 8DE
UK

Tel: +44 (0)1491 832111
Fax: +44 (0)1491 833508
Email: cabi@cabi.org
Web site: http: //www.cabi-publishing.org

CABI Publishing
10 E 40th Street
Suite 3203
New York, NY 10016
USA

Tel: +1 212 481 7018
Fax: +1 212 686 7993
Email: cabi-nao@cabi.org

A catalogue record for this book is available from the British Library, London, UK.

Library of Congress Cataloging-in-Publication Data

Tropical mycology / edited by Roy Watling ... [et al.].
p. cm.
Selected papers of the Millennium Symposium held April 2000 at the Liverpool John Moores University and organized by the British Mycological Society.
Includes bibliographical references and index.
Contents: v. 1. Macromycetes.
ISBN 0-85199-542-X (v. 1 : alk. paper)
1. Mycology--Tropics--Congresses. 2. Fungi--Tropics--Congresses. I. Watling, Roy.
QK615.7 .T76 2001
616.9′69′00913--dc21 2001025877

ISBN 0 85199 542 X

Typeset in 10/12pt Photina by Columns Design Ltd, Reading.
Printed and bound in the UK by Biddles Ltd, Guildford and King's Lynn.

Contents

Dedication

It is with great pleasure that this volume is dedicated to the late Professor Edred John Henry Corner (1906–1996) who made so many fundamental contributions to tropical mycology, especially in the study of tropical macromycetes. He was a Fellow of the Royal Society of London and an Honorary Member of both the British Mycological Society and Mycological Society of America. Among the many honours received, he was the first recipient of the de Bary Medal struck by the International Mycological Association in 1996. He presented the Benefactor's lecture at the Tropical Mycology Meeting held in Liverpool in 1992 and was a contributor to the British Mycological Society's publication *Aspects in Tropical Mycology*. He has been an inspiration to many of the researchers whose work is described herein and his publications will be baselines from which much future work will flourish.

A full acount of E.J.H. Corner, his mycology and publications, authored by R. Watling, will appear in a special issue of *Flora Malesiana* commemorating E.J.H. Corner and Benjamin Stone.

Roy Watling
January 2001

(See Watling, R. and Ginns, J. (1998) *Mycologia* 90(4) 732–737.)

Contributors

A.M. Ainsworth, Cubist Pharmaceuticals, 545 Ipswich Road, Slough, Berkshire SL1 4EQ, UK

Timothy J. Baroni, Department of Biological Sciences, SUNY Cortland, PO Box 2000, Cortland, NY 13045, USA

Bart Buyck, Muséum National d'Histoire Naturelle, Laboratoire de Cryptogamie, 12 Rue Buffon, F-75005 Paris, France

Raul Camacho, Facultad de Ciencias, Escuela Superior Politecnica de Chimborazo, Riobamba, Ecuador

Sharon A. Cantrell, Center for Forest Mycology Research, Forest Products Laboratory, USDA-Forest Service, PO Box 1377, Luquillo, PR 00773-1377, USA

Tun Tschu Chang, Division of Forest Protection, Taiwan Forestry Research Institute, 53 Nan-hai Road, Taipei 100, Taiwan

Siu Wai Chiu, Department of Biology, The Chinese University of Hong Kong, Shatin, Hong Kong SAR, China

Christine S. Evans, Fungal Biotechnology Group, University of Westminster, 115 New Cavendish Street, London W1M 8JS, UK

Juliet C. Frankland, Merlewood Research Station, Centre for Ecology and Hydrology, Grange-over-Sands, Cumbria LA11 6JU, UK

P.J. Fisher, School of Biological Sciences, University of Portsmouth, King Henry Building, King Henry 1st Street, Portsmouth PO1 2DY, UK

Laura Guzmán-Dávalos, Departamento de Botánica y Zoologia, Universidad de Guadalajara, Apdo. postal 1-139, Zapopan, Jalisco 45110, Mexico

Roy E. Halling, Institute of Systematic Botany, The New York Botanical Garden, Bronx, NY 10458-5126, USA

Marja Härkönen, Department of Ecology and Systematics, PO Box 7, FIN-00014, University of Helsinki, Finland

John N. Hedger, Fungal Biotechnology Group, University of Westminster, 115 New Cavendish Street, London W1M 8JS, UK

Susan Isaac, Department of Genetics and Microbiology, University of Liverpool, PO Box 147, Liverpool L69 3BX, UK

T.N. Kaul, 249 Nilgiri Apartments, Alaknanda, New Delhi 110 019, India

D. Jean Lodge, Center for Forest Mycology Research, Forest Products Laboratory, USDA-Forest Service, PO Box 1377, Luquillo, PR 00773-1377, USA

David Moore, School of Biological Sciences, University of Manchester, Oxford Road, Manchester M13 9PT, UK

Gregory M. Mueller, Department of Botany, The Field Museum of Natural History, Chicago, IL 60605-2496, USA

Maria Núñez, Mycoteam, Forskningsveien 3b, PO Box 5, Blindern, N-0313, Oslo, Norway

Clare H. Robinson, Division of Life Sciences, King's College, University of London, Franklin-Wilkins Building, 150 Stamford Street, London SE1 8WA, UK

See-Su Lee, Forest Research Institute, Malaysia, Kepong, 52109 Kuala Lumpur, Malaysia

Jogeir N. Stokland, Norwegian Institute of Land Inventory, PO Box 115, N-1430 Ås, Norway

D.J. Stradling, Hatherly Laboratories, Department of Biological Sciences, University of Exeter, Exeter EX4 4PS, UK

Suhirman, Indonesian Botanic Gardens, Jl. Juanda 13 Bogor 16122, Indonesia

Evelyn Turnbull, Royal Botanic Garden, Edinburgh EH3 5LR, UK

Millie A. Ullah, Fungal Biotechnology Group, University of Westminster, 115 New Cavendish Street, London W1M 8JS, UK

Anne-mieke Verbeken, Ghent University, Department of Biology, Ledeganckstraat 35, B-9000 Ghent, Belgium

Roy Watling, Caledonian Mycological Enterprises, Crelah, 26 Blinkbonny Avenue, Edinburgh EH4 3HU, UK and Royal Botanic Garden, Edinburgh EH3 5LR, UK

Preface

Half of the British Mycological Society's membership is from the British Isles, yet many are from countries which are generally considered to be situated in 'the Tropics'. Since the Sixth General Meeting of the Society held in Birmingham in 1988 members have recognized the increasing importance of tropical mycology and Council felt that it would be appropriate to bring both these factors of membership and research interests together for its millennium activities. A symposium was therefore organized in April 2000 to cover as many aspects as possible of the tropical mycobiota. This volume is one of two which result from this millennium symposium held at John Moores University, in Liverpool. In order to broaden the scope and the depth of the subjects covered in this book other chapters not formally presented at the symposium are included. This volume deals with the macrofungi; a sister volume deals with the microfungi.

The British Mycological Society has now organized two meetings specifically on Tropical Mycology and both were held in Liverpool, coincidentally a city with a long history of contacts with the area of the world which lies between 25° north and south of the Equator known to us all as 'the Tropics'. These books will therefore act as a companion to the proceedings of the meeting held in 1992 which was published 1 year later under the title *Aspects of Tropical Mycology*. Thoughts expressed therein have now been consolidated and the chapters included here are a summary of the state of knowledge in their respective fields. Some of the studies have indirect connections to the two expeditions the Society has organized to tropical areas, the first to Ecuador in 1993 and the second to Thailand in 1997.

In the selection of subjects, attempts have been made to bring together papers which offer a wide spectrum of information, linking results from the New

World and data from the Old World tropics. Contributions by 25 researchers cover in 12 chapters every major area of the tropics. Although our knowledge of the tropical mycotas is far from adequate, the surge of interest in the fungi of these areas has expanded our knowledge several-fold since 1992.

In this first volume chapters undoubtedly interlink, but they have been arranged in such a way as to cover the broader issues of ectomycorrhizal and floristic studies (Chapters 1–4), systematics (Chapter 5), lignicolous fungi (Chapters 6 and 7) and their enzymatic activity (Chapter 8), insect–fungal interrelationships (Chapter 9) and conservation and mycophagy (Chapters 10 and 11). The book finishes with a chapter on modern technologies that can be applied to the cultivation of edible fungi, the majority of which originate in the tropics.

It is well appreciated by researchers studying biodiversity that fungi play an important and integral part of any ecosystem and that mycodiversity studies are essential to our understanding of the world's natural heritage. It is therefore appropriate to include an account (Chapter 1) which focuses on the macrofungi of the oakwoods of the Neotropics, an area extremely rich in boletes and agarics but until recently little known mycologically. In Chapter 2 the fungi forming sheathing mycorrhizas in Africa are considered, a group of organisms which until relatively recently were not thought to be widespread in the tropics. The field results in this chapter complement the quantitative work described in Chapter 3, based on South-east Asian rainforest communities and where the analyses indicate that we should look much further in our search for potential ectomycorrhizal hosts. It is appropriate to include in this volume information (Chapter 4) on the ambitious Greater Antilles Project, which has involved so many macromycologists over the last few years and which is now entering an exciting analytical stage. One large group of basidiomycetes in the tropics is the dark-spored agarics, although in some parts of South-east Asia and in Africa genera such as *Cortinarius* and *Inocybe* are either absent or extremely rare. With particular emphasis on *Gymnopilus*, Chapter 5 deals with the dark-spored agarics of Mexico, a rather important centre of diversity of many organisms.

Much discussion in natural history circles has surrounded the presence or absence of a division separating the Australasian fauna and flora into two distinct groups, one to the south and one the north of an imaginary line. Chapter 6 questions whether such a division can be demonstrated for the distribution of bracket fungi or polypores. Chapter 7 explores the biology and ecological parameters of one member of this group of fungi, *Phellinus noxius*, a very important pathogen of the tropics. Tropical fungi, like their temperate counterparts, have a wide array of enzymes to break down organic material, either of standing trees or woody litter, and by doing so release essential nutrients for recycling. Chapter 8 looks at the production of lignolytic enzymes in various assemblages of fungi found in one area of the New World. There is also a complex series of enzymes, chemical messengers and ecological interactions operating in the amazing close relationship between the Neotropical leaf-cutting ants and their fungal associates. Chapter 9 examines this complex system under laboratory conditions in a way previously not undertaken.

Despite recent advances our knowledge of the world's mycota is still in a very meagre state. Concerning tropical constituents, of which there are undoubtedly a vast number of undescribed species, we sadly lack even fundamental information. Fungus lists and compilations are essential to stimulate the smallest glimmer of interest in fungal conservation and Chapter 10 develops this by trying to identify some of the issues which concern mycologists in India and how these are being tackled. Meanwhile the problems of the identity of those fungi which are being collected as food or medicine in China and East Africa, and the way the indigenous people identify the poisonous specimens, are addressed in Chapter 11. Examination of the development and physiology of tropical mushrooms in both the field and in the laboratory (Chapter 12) will lead to a better understanding, and possibly more efficient exploitation, of these fungi. The general public in Europe are blissfully unaware of the tropical origin of many of the fungi which are now appearing on supermarket shelves. An additional bonus of this book and its sister volume is the vast array of references which accompany each chapter; such references act in themselves as a valuable research tool.

It is hoped that this collection of chapters will stimulate many students of mycology and be helpful to a whole range of biologists interested in tropical biodiversity. It would be outstanding if they could also act as pump-primer to support research programmes and even a demand to launch a third British Mycological Society tropical expedition.

It is my great pleasure to thank an industrious team of editors, Martyn Ainsworth, Susan Isaac and Clare Robinson, who have given their all unselfishly since April 2000 in reviewing and editing the chapters for this publication and its sister volume. Special thanks to Juliet Frankland who has helped me finalize the drafts ready for technical adjustment and presentation so they could be brought together as soon after the meeting as possible in order to act as a baseline for future studies.

Roy Watling,
Caledonian Mycological Enterprises, Edinburgh, and
Royal Botanic Garden, Edinburgh, UK

Agarics and Boletes of Neotropical Oakwoods 1

R.E. Halling[1] and G.M. Mueller[2]

[1]Institute of Systematic Botany, The New York Botanical Garden, Bronx, NY 10458-5126, USA; [2]Department of Botany, The Field Museum of Natural History, Chicago, IL 60605-2496, USA

Introduction

The agarics and boletes in Neotropical forests exhibit several different patterns of origin and distribution. The patterns and diversity are directly related to the types of habitats in which these fungi occur and have been influenced by geological history, native and introduced phanerogamic vegetation, climate and, ultimately, human impact. Specifically, the aforementioned factors affect the distribution of those agarics and boletes that occur naturally in the montane oak forests of Costa Rica and Colombia. Much of the descriptive literature documenting the diversity of Neotropical, montane oak forest agarics and boletes has been cited previously (Halling and Franco-M., 1996; Halling and Mueller, 1999).

Others have speculated previously on the biogeographic connections of a few agaric genera and species (e.g. Guzmán, 1974; Moser and Horak, 1975; Halling and Ovrebo, 1987; Guzmán *et al.*, 1989; Mueller and Strack, 1992; Mueller and Halling, 1995; Halling, 1997a). Based on concentrated fieldwork in Central and South America over the last decade, with particular emphasis on ectomycorrhizal fungi, a mass of data has been compiled, the interpretation of which suggests clear biogeographic tendencies or connections.

Trappe (1977) postulated that the present disjunct distribution of many truffle-like fungi (Fig. 1.1) in Europe and North America, which are dependent on animals for dispersal, is best explained by their widespread occurrence throughout Laurasia prior to the tectonic events that broke up the animals' overland migration routes some 50 million years ago. While it is commonly accepted that above ground forms, like mushrooms, were ancestral to those that are truffle-like (e.g. Thiers, 1984; Bruns *et al.*, 1989; Mueller and Pine, 1994), those mushrooms that had the potential to form ectomycorrhizas would have had to have originated at some point before the 50 million year mark. Based on molecular clock dating of DNA sequences, Berbee and Taylor (1993) suggested that divergence of the homobasidiomycetes occurred in the Triassic, ~220 million years ago ± 50 million years. According to Newman and Reddell (1987), 95% of the *Pinaceae* form ectomycorrhizas. If we assume that the *Pinaceae* are monophyletic and plesiomorphic ectotrophs then *Compsostrobus*, a late Triassic fossil that is thought to be an ovulate cone of *Pinaceae* (Delevoryas and Hope, 1987), suggests that ectomycorrhizal homobasidiomycetes could have arisen by ~200 million years ago (Hibbett *et al.*, 1997). Ectomycorrhizal fungi probably diversified later in the Jurassic (208–146 million years ago), when ectomycorrhizal gymnosperms had become globally established before the breakup of Pangaea. The subsequent diversification of angiosperms was probably a catalyst for a boost in the diversity of ectomycorrhizal fungi soon after these plants appeared in the Cretaceous (146–165 million years ago) (Raven and Axelrod, 1974; Talent, 1984; Truswell *et al.*, 1987).

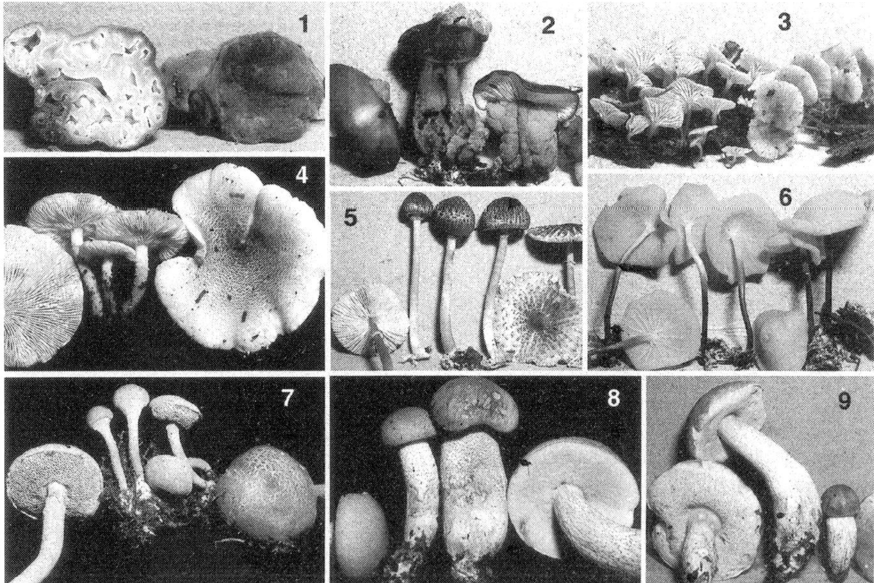

Fig. 1.1. *Hydnotrya tulasnei* (× 0.58). **Fig. 1.2.** *Gymnopus nubicola* with *Syzygospora* parasite (× 0.39). **Fig. 1.3.** *Xeromphalina kauffmanii* (× 0.77). **Fig. 1.4.** *Ripartitella brasilensis* (× 0.39). **Fig. 1.5.** *Rugosospora pseudorubiginosa* (× 0.39). **Fig. 1.6.** *Hymenogloea papyracea* (× 0.19). **Fig. 1.7.** *Pulveroboletus ravenellii* (× 0.58). **Fig. 1.8.** *Tylopilus chromapes* (× 0.19). **Fig. 1.9.** *Tylopilus cartagoensis* (× 0.19).

Stratigraphic evidence in Mesoamerica suggests that the closure of the Panamanian Isthmus occurred from about 3.5–2.4 million years ago in the mid-Pliocene (Marshall and Sempere, 1993; Graham, 1995) allowing for biotic interchange in both directions across latitudes between North and South America. The closure of this portal becomes relevant in the discussion of the recent age of oak forests in South America.

Since the mid to late 1950s, after centuries of deforestation, many areas of the montane Neotropics and elsewhere in the temperate zones were reforested with plantations of exotic trees, namely *Pinus* and *Eucalyptus* spp. The commercial nursery stock of these trees was apparently inoculated with only a few mutualistic mycorrhizal fungi, *Amanita muscaria* (L. : Fr.) Hook. and *Suillus luteus* (L. : Fr.) Gray in particular for pines, and *Laccaria fraterna* (Cooke & Massee: Sacc.) Pegler and *Hydnangium carneum* Wallr. apud Klotzsch for *Eucalyptus*. For example, the Cotopaxi volcano in Ecuador harbours an exotic Monterey pine forest which produces, from personal observation, fruitings of *Suillus luteus* all the year round, to the exclusion of other macrofungi. It is interesting to note that in its native range in California, Monterey pine does not have *S. luteus* as a mycorrhizal symbiont as observed by the authors and H.D. Thiers (California, 1984, personal communication).

In native montane habitats such as the páramo regions above the timberline exemplified by a zone around the Chimborazo volcano in Ecuador, where frost may occur any day of the year, saprobic agarics occur that have north temperate affinities, for example: *Leptonia serrulata* (Fr.) Kummer, and a relative of *Gymnopus dryophilus* (Bull. : Fr.) Murr., named *Gymnopus nubicola* Halling. In its migration from the north, the latter was accompanied by its obligate north temperate parasite, *Syzygospora* (Fig. 1.2). In this zone above the timberline, there are no ectomycorrhizal agarics.

Other non-Andean, montane habitats include the harsh environment of the sandstone table mountains (tepuis) of the Guyana Highlands in north-eastern South America (like Mt Roraima, which inspired A. Conan Doyle's *The Lost World* (1912)). The bizarre vascular plant flora was first explored by the Schomburgk brothers and Everard Im Thurn, and has been extensively documented by Basset Maguire and collaborators since the mid-1950s (Maguire *et al.*, 1953). These regions have been little explored by mycologists. The summits of these tropical plateaux are often rocky, or dominated by open, extremely wet, savanna type vegetation with a low pH. However, in areas inhabited by gallery forest where leaf litter or dead wood accumulates, *Xeromphalina nubium* Redhead & Halling (1987) occurs whose sister species appears to be *Xeromphalina kauffmanii* A.H. Sm. (Fig. 1.3), a North American agaric that grows on oak logs and has a southern distribution limit in the oak forests of montane Panama.

On the other hand, agarics of lowland rainforests exhibit widespread to restricted patterns of distribution but appear not to have species overlap with the temperate zones. Examples here include the pantropical *Ripartitella brasiliensis* (Speg.) Singer (Fig. 1.4), and the amphiatlantic genus *Rugosospora* (*R.*

pseudorubiginosa (Cifuentes & Guzmán) Guzmán & Bandala) (Fig. 1.5), whose Neotropical existence was first noted by Cifuentes and Guzmán (1981). A widespread Neotropical example is provided by *Trogia cantharelloides* (Mont.) Pat. Apparent endemics of lowland to submontane Central America and South America comprise the marasmioid *Hymenogloea papyracea* (Berk & M.A. Curtis) Singer (Fig. 1.6) and the recently described *Marasmiellus volvatus* Singer. However, the known distribution of the latter may just be a reflection of where it has been encountered in under-collected habitats by fortunate, knowledgeable agaricologists.

In montane ectomycorrhizal habitats, Neotropical oak forests are narrowly restricted to the main cordilleras of Mexico, southwards through the Andes to southern Colombia. Dating of pollen in core samples in the vicinity of Bogotá, places their appearance at ± 340,000 years ago (Hooghiemstra and Cleef, 1995). These fossil data indicate that oak is a recent immigrant to central and northern South America. Oaks are abundant and speciose in North America and Mexico, but in Costa Rica there are just ± 12 species (Burger, 1977), whereas in Colombia there is only one. In Costa Rica, according to Kappelle *et al.* (1992), 80% of the canopy and 95% of the basal area of montane forests in the Cordillera Talamanca consist of oaks, whereas 75% of all other plant genera in the understorey and subcanopy are of tropical origin. The agarics and boletes found in these forests show definite affinities with North Temperate taxa and they undoubtedly migrated southward from Mexico with those forest communities.

Documentation and personal observations of ectomycorrhizal associations and distributions during the last two decades have revealed some distinctive patterns: (i) relictual disjunct distributions; (ii) generic and specific distributions along a cline; (iii) local endemism; (iv) high generic similarity; (v) low species similarity; and (vi) high levels of macromycete diversity. These patterns are discussed below in more detail.

Relictual Disjunct Distributions

Based on its current distribution, *Pulveroboletus ravenellii* (Berk. & M.A. Curtis) Murr. (Fig. 1.7), in its many morphs, exhibits a relictual disjunct distribution (personal observation). It is likely to have an ancient genome with minor phenotypic change, and is a generalist in its mycorrhizal associations. It partners with *Pinaceae* and various dicotyledonous genera in eastern and western North America. In its southern distribution in the Americas, *P. ravenellii* associates solely with *Quercus*. It is also reported from *Fagaceae* forests in south-eastern Asia (Corner, 1972), *Myrtaceae* and *Casuarinaceae* in Queensland, Australia (personal observation), but it is notably absent in Europe.

Tylopilus subgenus *Roseoscabra* (typified by *Tylopilus chromapes* (Frost) A.H. Sm. & Thiers fide Wolfe and Bougher, 1993) (Fig. 1.8) presents a slightly different scenario. Wolfe and Bougher (1993) added eight species to the original *T.*

chromapes, including *Tylopilus cartagoensis* C.B. Wolfe & Bougher (Fig. 1.9) from Costa Rica and *Tylopilus queenslandianus* C.B. Wolfe & Bougher from Australia among others. These same authors speculated that there was a Laurasian origin of the ancient genotype, and subsequent migration to and speciation in Australia during Pleistocene glaciations via land bridges. Today known associates are *Pinaceae*, *Fagaceae*, *Betulaceae*, *Salicaceae*, *Myrtaceae*, *Mimosaceae*, and *Casuarinaceae* (known distribution areas of the subgenus are: eastern USA, Costa Rica, Japan to China to north-east Australia but not in Europe). Again, subgenus *Roseoscabra* probably represents an ancient genome, with a generalist mycorrhizal proclivity, but with some phenotypic change or genetic drift/allopatric speciation.

Rozites is an obligately ectomycorrhizal agaric with three known species in the Northern Hemisphere (*Rozites caperata* (Pers. : Fr.) P. Karst. and the recently described *Rozites colombiana* Halling & Ovrebo, and the locally restricted *Rozites emodensis* (Berk.) Moser from the Himalayas). Recently, Bougher *et al.* (1994) indicated that *Rozites* has essentially a Gondwanan distribution and appears to have co-evolved with a Cretaceous fagalean complex with more genetic diversity going south (18 species known today), where it co-evolved with *Nothofagus* (Australia, New Zealand, Chile, Argentina, New Caledonia, Papua New Guinea) and/or with *Myrtaceae* (in Australia). Lesser diversity clearly evolved in the Northern Hemisphere (only three species) after continental breakup. We know that *Nothofagus* was distributed across southern Australia when it rafted away from Antarctica, but it gradually became extinct in part of its range with increasing aridity across in the region known as the Nullarbor Plain. Fossils of *Nothofagus* on that continent and its existence in South America and elsewhere have been the subject of much biogeographical debate (references cited in Bougher *et al.*, 1994). When *Nothofagus* became extinct in south-western Australia in the Eocene/Pliocene, Bougher *et al.* (1994) speculated that the relictual *Rozites symeae* Bougher, Fuhrer, & Horak apparently had the capacity to switch partners from *Nothofagus* to species of *Myrtaceae*. Those authors also pointed out that there is another species, *Rozites fusipes* Horak & Taylor (apparently a sister taxon to *R. symeae*), that is associated with *Nothofagus* in New Zealand and with *Nothofagus* and *Eucalyptus* in Tasmania. Bougher (1987) has demonstrated pure culture synthesis of ectomycorrhizas with *Nothofagus* and *Eucalyptus* using isolates of *Descolea maculata* Bougher & Malajczuk from Western Australia eucalypt forests. In the New World, Halling (1989) and later Mueller and Strack (1992) documented similar host shifts of mushroom species from conifers in the United States to oaks in Colombia and Costa Rica.

North/South Clinal Distribution

A general analysis of phenetic similarities in one family and three genera of ectomycorrhizal mushrooms showed greater (at higher taxonomic levels) or lesser (at lower taxonomic levels) affinities with North Temperate taxa of the

Western Hemisphere (Mueller and Halling, 1995). These latter data indicate a definite north/south clinal distribution pattern. For example, species with a distribution range from the North Temperate Zone to southern Colombia would comprise: *Cortinarius iodes* Berk. & M.A. Curtis, *Lactarius indigo* (Schwein.) Fr., *Lactarius atroviridis* Peck (Fig. 1.10), *Laccaria amethystina* Cooke, and *Strobilomyces confusus* Singer. In these particular examples (except for *L. amethystina*), the north temperate distribution is restricted to eastern North America. These taxa exhibit the largest distribution range.

Another clinal pattern describes species with a north temperate to Costa Rica distribution. In this case, one would encounter *Asterophora parasitica* (Fr.) Singer, *Boletus frostii* J.L. Russell (Fig. 1.11), *Lactarius psammicola* A.H. Sm., *Lactarius rimosellus* Peck, *Russula nigricans* (Bull. : Fr.) Fr., and *Tylopilus eximius* (Peck) Singer. While *Asterophora* is an obligate parasite of the obligately ectomycorrhizal *Russula* and widely distributed in the Northern Hemisphere (as is *R. nigricans*), the other species are part of the eastern North American mycota.

Local Endemism

In the montane Neotropics of Central and northern South America, observations show that there are degrees of apparent endemism with species ranging from locally distributed to quite restricted. Some Neotropical oak endemics are *Amanita garabitoana* nom. prov., *Laccaria gomezii* G.M. Mueller & Singer (Fig. 1.12), *Lactarius costaricensis* Singer, *Leccinum andinum* Halling, and *R. colombiana*. Here, it is important to note that the genera are widespread north

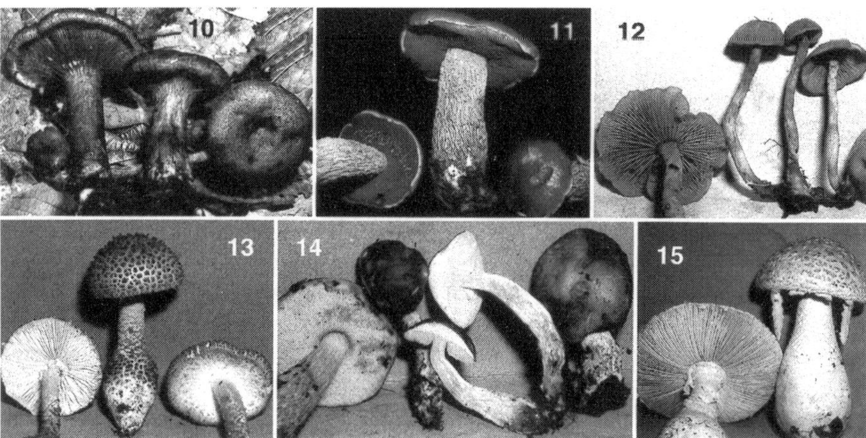

Fig. 1.10. *Lactarius atroviridis* (× 0.39). **Fig. 1.11.** *Boletus frostii* (× 0.19). **Fig. 1.12.** *Laccaria gomezii* (× 0.39). **Fig. 1.13.** *Amanita costaricensis* nom. prov. (× 0.10). **Fig. 1.14.** *Leccinum talamancae* (× 0.19). **Fig. 1.15.** *Amanita conara* nom. prov. (× 0.10).

temperate entities (except see discussion of *Rozites* above), but the species appear endemic to the montane Neotropics. Currently, in the case of more restricted endemics, one would encounter *Amanita costaricensis* nom. prov. (Fig. 1.13), *Leccinum talamancae* Halling, Gómez, & Lannoy (Fig. 1.14), *Amanita conara* nom. prov. (Fig. 1.15), *Boleus flavoniger* Halling, G.M. Mueller, & Gómez, and *Tricholosporum violaceum* Halling & Franco-M. As mentioned above, the genera are widespread north temperate taxa, but the species appear to be locally restricted to only a few sites in the Cordillera Talamanca of Costa Rica.

Generic Similarity

The *Boletaceae* are a conspicuous element in any north temperate woodland ecosystem. Oak forests of the montane Neotropics are no exception. In an analysis of 24 bolete genera *sensu* Singer (1986), which are almost always ectomycorrhizal, 18 occur in Costa Rica and Colombia (Neotropical oak forests) with a relative similarity of ± 94% with the North Temperate Zone (cf. Mueller and Halling, 1995; Halling, 1997b). These data would add support to the aforementioned conclusions that ectomycorrhizal plants (like oaks) migrated from the Northern Hemisphere. In addition, it appears that whole communities migrated (plants and mycorrhizal fungi, as well as obligate parasites of cited fungi).

Species Similarity

Halling (1997b) provided evidence that the overlap of species similarity of selected genera of boletes with the North Temperate Zone was low, ranging from one-fifth to less than one-half. However, not mentioned in that publication is the fact that Neotropical bolete species are most closely allied with species from eastern North America (e.g. *B. flavoniger* is an ally with *Boletus ornatipes* Peck and *Boletus retipes* Berk. & M.A. Curtis (Halling, 1997b)).

Macromycete Diversity

In a survey of 0.1 ha (1000 m^2) in the Cordillera Talamanca, Costa Rica (G.M. Mueller and R.E. Halling, Chicago and New York, unpublished data), there are over 200 species of macrofungi in a parcel of native forest that includes only 20–25 species of vascular plants (M. Kappelle, Costa Rica, 1997, personal communication), and just two species of ectomycorrhizal angiosperms (*Quercus seemannii* Liebm. and *Quercus copeyensis* C.H. Muller). Such numbers lend credence to Hawksworth's (1991) hypothesis concerning the world's diversity of fungi. Additional unpublished data from further collecting in Neotropical habitats bolster Hawksworth's contentions.

Conclusions

In summary, apparent restricted distributions of agarics and boletes will always be discovered based on the work of those who frequent under-collected habitats. Relictual disjunct distributions reflect original widespread occurrence of old genomes before continental breakup with little subsequent phenotypic variation. The taxonomic affinities of non-montane, lowland agarics and boletes range from restricted endemics to those with widespread pantropical occurrence. At this time, little can be said about non-Andean, montane agaric biogeography of the Neotropics; few mycologists have collected in those remote sites in the last millennium. Neotropical montane habitats dominated by oak decidedly harbour agarics and boletes with genera and species whose genealogy is derived from the North Temperate Zone. These taxa exhibit a north to south clinal distribution pattern. While generic similarity is high and species similarity low in Neotropical montane oak forests, the expectation is that these trends will remain even as additional diversity is discovered.

Acknowledgements

Much of our work has been supported by the National Science Foundation, BSR 86-00424 and DEB-9972018 (REH), BSR 86-07106 and DEB-9972027 (GMM), and by the Office of Forestry, Environment and Natural Resources, Bureau of Science and Technology, of the US Agency for International Development in NSF Grant No. DEB-9300798. The advice and aid of our many collaborators in Latin America and North America are very gratefully acknowledged. The senior author is grateful to the British Mycological Society for the invitation to, and waiver of some expenses at, the Tropical Mycology 2000 Symposium in Liverpool.

References

Berbee, M.L. and Taylor, J.W. (1993) Dating the evolutionary radiations of the true fungi. *Canadian Journal of Botany* 71, 1114–1127.

Bougher, N.L. (1987) The systematic position and ectomycorrhizal status of the fungal genus *Descolea*. PhD thesis, University of Western Australia, Perth, Australia.

Bougher, N.L., Fuhrer, B.A. and Horak, E. (1994) Taxonomy and biogeography of Australian *Rozites* species mycorrhizal with *Nothofagus* and Myrtaceae. *Australian Systematic Botany* 7, 353–375.

Bruns, T.D., Fogel, R., White, T.J. and Palmer, J.D. (1989) Accelerated evolution of a false-truffle from a mushroom ancestor. *Nature* 339, 140–142.

Burger, W. (1977) Flora Costaricensis: Fagaceae. In: Burger, W. (ed.) *Fieldiana* Botany Series 40, pp. 59–80.

Cifuentes, J. and Guzmán, G. (1981) Descripción y distribución de hongos tropicales (Agaricales) no conocidos previamente en México. *Boletin de la Sociedad Méxicana de Micologia* 16, 35–61.

Corner, E.J.H. (1972) *Boletus in Malaysia.* Government Printing Office, Singapore.

Delevoryas, T. and Hope, R.C. (1987) Further observations on the late Triassic conifers *Compsostrobus neotericus* and *Voltzia andrewsii. Review of Paleobotany and Palynology* 51, 59–64.

Doyle, A.C. (1912) *The Lost World.* Hodder & Stoughton, George H. Doran Co., New York.

Graham, A. (1995) Development of affinities between Mexican/Central American and northern South American lowland and lower montane vegetation during the Tertiary. In: Churchill, S., Balslev, H., Forero, E. and Luteyn, J. (eds) *Biodiversity and Conservation of Neotropical Montane Forests.* New York Botanical Garden Press, New York, pp. 11–22.

Guzmán, G. (1974) El género *Fistulinella* Henn. (=*Ixechinus* Heim) y las relaciones florísticas entre México y Africa. *Boletin de la Sociedad Méxicana de Micologia* 8, 53–63.

Guzmán, G., Bandala, V., Montoya, L. and Saldarriaga, Y. (1989) Nuevas evidencias sobre las relaciones micofloristicas entre Africa y el Neotropico. El genero *Rugosospora* Heinem. (Fungi, Agaricales). *Brenesia* 32, 107–112.

Halling, R.E. (1989) A synopsis of Colombian boletes. *Mycotaxon* 34, 93–113.

Halling, R.E. (1997a) Notes on *Collybia* V. *Gymnopus* section *Levipedes* in tropical South America, with comments on *Collybia. Brittonia* 48, 487–494.

Halling, R.E. (1997b) Boletaceae (Agaricales): Latitudinal biodiversity and biological interactions in Costa Rica and Colombia. *Revista Biologia Tropical* 44 (supple. 4), 111–114.

Halling, R.E. and Franco-M., A.E. (1996) Agaricales from Costa Rica: new taxa with ornamented spores. *Mycologia* 88, 666–670.

Halling, R.E. and Mueller, G.M. (1999) New boletes from Costa Rica. *Mycologia* 91, 893–899.

Halling, R.E. and Ovrebo, C.L. (1987) A new species of *Rozites* from oak forests of Colombia with notes on biogeography. *Mycologia* 79, 674–678.

Hawksworth, D.L. (1991) The fungal dimension of biodiversity: magnitude, significance, and conservation. *Mycological Research* 95, 641–655.

Hibbett, D.S., Grimaldi, D. and Donoghue, M.J. (1997) Fossil mushrooms from Miocene and Cretaceous ambers and the evolution of Homobasidiomycetes. *American Journal of Botany* 84, 981–991.

Hooghiemstra, H. and Cleef, A.M. (1995) Pleistocene climatic change and environmental and generic dynamics in the north Andean montane forest and páramo. In: Churchill, S., Balslev, H., Forero, E. and Luteyn, J. (eds) *Biodiversity and Conservation of Neotropical Montane Forests.* New York Botanical Garden Press, New York, pp. 35–49.

Kappelle, M., Cleef, A.M. and Chaverri, A. (1992) Phytogeography of Talamanca montane *Quercus* forests, Costa Rica. *Journal of Biogeography* 19, 299–315.

Maguire, B., Cowan, R.S. and Wurdack, J.J. (1953) The botany of the Guayana Highland. *Memoirs of the New York Botanical Garden* 8, 87–160.

Marshall, L.G. and Sempere, T. (1993) Evolution of the Neotropical Cenozoic land mammal fauna in its geochronologic, stratigraphic and tectonic context. In: Goldblatt, P. (ed.) *Biological Relationships between Africa and South America.* Yale University Press, New Haven, Connecticut, pp. 329–392.

Moser, M. and Horak, E. (1975) *Cortinarius* Fries und nahe verwandte Gattungen in Südamerika. *Beihefte zur Nova Hedwigia* 52, 1–628.

Mueller, G.M. and Halling, R.E. (1995) Evidence for high biodiversity of Agaricales (Fungi) in Neotropical montane *Quercus* forests. In: Churchill, S., Balslev, H., Forero, E. and Luteyn, J. (eds) *Biodiversity and Conservation of Neotropical Montane Forests.* New York Botanical Garden Press, New York, pp. 303–312.

Mueller, G.M. and Pine, E.M. (1994) DNA data provide evidence on the evolutionary relationships between mushrooms and false truffles. *McIlvainea* 11, 61–74.

Mueller, G.M. and Strack, B.A. (1992) Evidence for a mycorrhizal host shift during migration of *Laccaria trichodermophora* and other agarics into Neotropical oak forests. *Mycotaxon* 45, 249–256.

Newman, E.I. and Reddell, P. (1987) The distribution of mycorrhizas among families of vascular plants. *New Phytologist* 106, 745–751.

Raven, P.H. and Axelrod, D.A. (1974) Angiosperm biogeography and past continental movements. *Annals of the Missouri Botanical Garden* 61, 539–673.

Redhead, S.A. and Halling, R.E. (1987) *Xeromphalina nubium* sp. nov. (Basidiomycetes) from Cerro de la Neblina, Venezuela. *Mycologia* 79, 383–386.

Singer, R. (1986) *The Agaricales in Modern Taxonomy*, 4th edn. Koeltz Scientific Books, Koenigstein, Germany.

Talent, J.A. (1984) Australian biogeography past and present: determinants and implications. In: Veevers, J.J. (ed.) *Phanerozoic Earth History of Australia.* Clarendon Press, Oxford, pp. 57–93.

Thiers, H.D. (1984) The secotioid syndrome. *Mycologia* 76, 1–8.

Trappe, J.M. (1977) Biogeography of hypogeous fungi: trees, mammals, and continental drift. *Abstracts 2nd International Mycological Congress*, p. 675.

Truswell, E.M., Kershaw, A.P.J. and Sluiter, I.R. (1987) The Australian-southeast Asian connection: evidence from the paleobotanical record. In: Whitmore, T.C. (ed.) *Biogeographical Evolution of the Malay Archipelago.* Oxford University Press, Oxford, pp. 32–49.

Wolfe, C.B. Jr and Bougher, N.L. (1993) Systematics, mycogeography, and evolutionary history of *Tylopilus* subg. *Roseoscabra* in Australia elucidated by comparison with Asian and American species. *Australian Systematic Botany* 6, 187–213.

Diversity and Ecology of Tropical Ectomycorrhizal Fungi in Africa 2

A. VERBEKEN[1] AND B. BUYCK[2]

[1]*Ghent University, Department of Biology, Ledeganckstraat 35, B-9000 Ghent, Belgium;* [2]*Muséum National d'Histoire Naturelle, Laboratoire de Cryptogamie, 12 Rue Buffon, F-75005 Paris, France*

Introduction

Contrary to general opinion some 20 years ago, ectomycorrhizal symbiosis is now considered to be very important in African ecosystems (e.g. Högberg and Piearce, 1986; Connel and Lowman, 1989; Buyck, 1994b). Ectomycorrhizas were demonstrated for the first time on roots of *Gilbertiodendron dewevrei* (De Wild.) Léonard, an important canopy tree in the Guineo-Congolian rainforest (Peyronel and Fassi, 1957). Later, ectomycorrhizas were also discovered on several other tree species and genera, including many characteristic miombo wood trees with *Brachystegia* spp. as the main component (Högberg and Nylund, 1981; Högberg, 1982; Högberg and Piearce, 1986; Thoen, 1993). Since then the study of ectomycorrhizas in tropical Africa has gained much interest but has mostly concentrated on the host trees. A review of the history of ectomycorrhizal research in tropical Africa is given by Fassi and Moser (1991).

Although the taxonomic knowledge of these fungi is very incomplete and their ecology has not yet been studied, some preliminary observations on the taxonomic diversity, ectomycorrhizal status, host specificity, geographic distribution, ecological range and phenology can be made.

The observations presented here were mainly based on a detailed revision of tropical African *Lactarius* by the first author, which involved about 1000 collections from 130 sites in 18 countries. An important part of the collecting was undertaken by the second author in miombo woodlands in northern Zambia and Burundi (Rumonge district). Extrapolations to other genera and quantitative estimations were based principally on unpublished field observations and existing collections of still undescribed species and, to a lesser degree, on literature data.

Taxonomic Diversity of the Ectomycorrhizal Fungi

The ectomycorrhizal fungi in tropical Africa are represented principally by *Amanita* Pers., *Cantharellaceae*, *Russulaceae* and *Boletales*, and to a lesser degree by hypogeous fungi, gasteromycetes, *Cortinariaceae* and *Gomphales* (Thoen and Bâ, 1989; Thoen and Ducousso, 1989; Buyck *et al.*, 1996; personal observations). Taxonomic knowledge of these groups is incomplete but advancing.

The genus *Lactarius* presently contains 83 accepted species and two varieties in tropical Africa (Karhula *et al.*, 1998; Verbeken, 1995, 1996a,b,c,d, 1998a,b, 2000; Verbeken *et al.*, 2000; Verbeken and Walleyn, 2000), but it is estimated that the total number of *Lactarius* in the African tropics exceeds 150 species. The number of undescribed collections in examined herbaria suggests at least 60 *Lactarius* species for the Zambezian and Sudanian woodlands alone. A similar number of *Lactarius* species is likely to be found in African, dense forest types. Taking into account that, even for a well-explored continent such as Europe, 21 new *Lactarius* species have been described since 1980, the above estimate can still be considered to be very conservative.

Table 2.1 compares the *Russula–Lactarius* species ratio (R/L) from a number of areas both in and outside Africa. Although the areas considered are very different in scale and exploration rates, the ratio (between 2.0 and 2.5) seems to be fairly constant and independent of continent or vegetation type. The total number of *Russula* species in tropical Africa (see Table 2.2) is estimated to range between 285 and 350 species so, taking into account the estimated number of *Lactarius* species, at least 400–500 species of *Russulaceae* are involved in ecto-

Table 2.1. A comparison between described *Russula* (R) and *Lactarius* (L) species in various parts of the world.

Area studied	No. of R	No. of L	R:L	Source
Europe	± 350	± 150	± 2.3	Courtecuisse and Duhem (1994)
Tropical Africa	165	85	1.9	Table 2.2
Germany	154	81	1.9	Krieglsteiner (1991)
Netherlands	118–128	56	2.1–2.3	Arnolds *et al.* (1995)
African rainforests	110	41	2.7	Buyck (1993,1994a, 1997); Verbeken (1996d)
South-east England	107	60	1.8	Dennis (1995)
Melle (Belgium), Geerbos	37	15	2.5	R. Walleyn (Gent, 1999, personal communication)
Australia	37	15	2.5	May and Wood (1997)
Madagascar	27	12	2.3	Heim (1938)
Shetland Isles[a]	9–10	3	3–3.3	Watling (1992)
Burundi, Rumonge	69%	31%	2.2	Buyck and Verbeken (1994)

[a]Introduced species excluded.

mycorrhizal symbioses in the vast forest and woodland areas throughout tropical Africa. The number of *Russulaceae* from tropical Africa, then, more or less equals that described from Europe.

Table 2.2 compares the number of described taxa accepted and the estimated total number of taxa for the other groups of ectomycorrhizal fungi in tropical Africa. From these data it seems that at least 400 ectomycorrhizal species remain to be described. This relatively low number can be explained by the paucity of ectomycorrhizal *Cortinariaceae* and *Tricholomataceae* in Africa; both families account for several thousands of species in the Northern Hemisphere.

Endemism of Tropical Ectomycorrhizal Fungi in Africa

None of the *Lactarius* and *Russula* species from indigenous, African forests and woodlands has been found elsewhere in the world. Earlier reports of European

Table 2.2. Main putative ectomycorrhizal fungi (EM) of tropical Africa.

	No. of taxa identified	Estimated no. taxa	Main references
Cantharellaceae	35–40	> 60	Heinemann (1958, 1966a); Eyssartier and Buyck (1999a,b)
Lactarius	85	> 125	Heim (1938, 1955); Verbeken (1995, 1996a,b,c,d, 1998a,b); Verbeken and Walleyn (2000); Verbeken *et al.* (2000); Karhula *et al.* (1998)
Russula	± 165	> 285	Heim (1938); Buyck (1993, 1994a, 1997, 1999); Härkönen *et al.* (1993)
Boletales[a]	140–145	> 200	Heinemann (1954, 1966b); Heinemann and Rammeloo (1980, 1983, 1986, 1987, 1989a,b); Watling and Turnbull (1993, 1994)
Amanita[a]	50–55	> 70	Beeli (1935); Härkönen *et al.* (1994); Pegler and Shah-Smith (1997); Walleyn and Verbeken (1999)
Cortinarius	1	> 10	Beeli (1928)
Inocybe	8	> 25	Pegler and Rayner (1969); Buyck and Eyssartier (1999)
Scleroderma	8	± 10	Dissing and Lange (1962); Demoulin and Dring (1971, 1975)
Sequestrate fungi	11	> 15	Castellano *et al.* (2000)
Total	± 410[b]	> 800	

[a]Probably including some non-mycorrhizal species.

[b]Excluding miscellaneous putative EM fungi (e.g. *Terfezia*, *Thelephoraceae* and *Coltricia*).

species were found to be misidentifications. We can, therefore, conclude that, to the best of our knowledge, the *Lactarius* and *Russula* mycota of tropical Africa is unique and completely endemic. This mycota is apparently also incompatible with exotic tree plantations, as only introduced fungi fruit in association with these foreign hosts; for example, *Russula pectinatoides* Peck (a temperate species) in Tanzanian pine plantations (Maghembe and Redhead, 1980; Härkönen *et al.*, 1993). An equally high degree of endemism is observed in most other groups of African ectomycorrhizal fungi (e.g. Robyns, 1935–1972; Heinemann and Rammeloo, 1972–1997; Buyck and Eyssartier, 1999) with the exception of *Scleroderma*, most reported species of which have a pantropical or almost cosmopolitan distribution. Interestingly, the few species reported from both tropical Africa and another continent, such as *Amanita hemibapha* (Berk. & Br.) Sacc., *Craterellus aureus* Berk. & M.A. Curtis or *Rubinoboletus balloui* (Peck) Heinem. & Rammeloo, belong to taxonomically critical complexes.

Some species have been recorded from both Madagascar and the African continent (e.g. *Lactarius rubroviolascens* R. Heim, *Russula gossypina* Buyck), but from the data available it is difficult to compare the ectomycorrhizal mycobiota of both areas, which have no host tree species in common.

On the other hand, close systematic affinities with species from other continents are obvious at the infrageneric level. In the case of *Lactarius*, examples are sections *Lactariopsidei* (tropical America), *Chamaeleontini* (South-east Asia), *Plinthogali* (Europe, North America, Asia) and *Rugati* (North Africa, Europe, Asia, North America, tropical America). The occurrence of the conspicuous secondary angiocarpic *Lactarius* species (section *Lactariopsidei*) in both tropical Africa and America is one of several resemblances observed between the mycobiota of these two regions (e.g. Guzmán *et al.*, 1989), suggesting an earlier link between the distributions of some fungi and some angiosperm genera (Watling, 1993).

The early isolated position of the African continent has undoubtedly resulted in several centres of diversification of tropical ectomycorrhizal fungi. The deserts in the north are a natural barrier for the migration of most ectomycorrhizal fungi and apparently no, or very few, indigenous ectomycorrhizal host plants occur in southern Africa (e.g. see Allsopp and Stock, 1993). All South African *Lactarius* and *Russula* species are associated with introduced *Pinus* (Doidge, 1950; Pearson, 1950; van der Westhuizen and Eicker, 1987, 1994) and are likely to be introduced along with their host (including *Russula capensis* A. Pearson, which is considered, most probably, to be identical to a European or North American species).

Nevertheless, indigenous ectomycorrhizal fungi might be present in southern Africa as some members of traditionally presumed ectomycorrhizal genera (*Amanita*, *Phlebopus*) occur in natural vegetation types in South Africa (Baxter, 1990; Reid and Eicker, 1991; van der Westhuizen and Eicker, 1994). The nutritional mode of such fungi needs investigation: a mycorrhizal relationship with herbaceous plants cannot be excluded, as has been shown for *Terfezia pfeilii* Henn. in Botswana (Taylor *et al.*, 1995). Furthermore, *Amanita foetidissima* D.A. Reid & Eicker, reported from Zambia (Pegler and Shah-Smith, 1997), is described as forming fairy rings in grasslands, away from any potential host tree (Reid and Eicker, 1991), and several *Phlebopus* typically sporulate in cultivated land or garden lawns.

Do All Tropical African Putative Ectomycorrhizal Species form Ectomycorrhizas?

Several species in these genera have sometimes been collected on rotten wood in various stages of decay. Basidiomes of *Lactarius gymnocarpus* R. Heim ex Singer were observed even 1 m up large tree stems in Korup National Park, Cameroon (R. Watling, Edinburgh, UK, 1994, personal communication). Sometimes these fungi with a 'lignicolous' habit are considered to be (facultative) saprotrophs (Singer, 1986). However, both Thoen and Watling independently observed mycelial links of 'lignicolous' *L. gymnocarpus* basidiomes with tree rootlets (R. Watling, Edinburgh, UK, 1994, personal communication; D. Thoen, Brussels, Belgium, 1998, personal communication). This 'lignicolous' habit is not restricted to the African tropics and has also been observed in tropical America (e.g. *Lactarius neotropicus* Singer: Pegler and Fiard, 1979), North America (e.g. *Lactarius vietus* (Fr.: Fr.) Fr., *Lactarius louisii* Homola : Hesler and Smith, 1979), Europe (e.g. *Lactarius camphoratus* (Bull.: Fr.) Fr.: Bresinsky and Stangl, 1970) and Asia (e.g. *Lactarius lignicola* W.F. Chiu: Chiu, 1945). Unpublished observations of the authors indicate a similar phenomenon in *L. vietus*, *L. camphoratus*, *Lactarius tabidus* Fr. and *Lactarius glyciosmus* (Fr.: Fr.) Fr.. As Hesler and Smith (1979) stated, 'it is common knowledge that the rootlets of trees often penetrate wet rotten logs and there show a profuse formation of mycorrhiza. The adaptation is one favouring both adequate moisture and air for the life processes of both host and fungus, and the basidiocarps near this centre of activity'. The 'lignicolous' habit is generally observed where the fitness of the ectomycorrhizal nutritional mode becomes more critical such as in very wet and poor soils (e.g. in swamp borders and riparian forest), burnt soils and rainforests (where the competition for nutrients is higher). This niche above ground level is also suggested as an adaptation to seasonally waterlogged sites (Singer and Araujo, 1986). African *Lactarius* species fruiting on wood have only been observed in dense forest, and not in other woodland types; indeed the 'lignicolous' habit is not a constant characteristic of most species studied in this respect. In African *Lactariopsis* and South American and Asian *Pleurogala* this character was over-emphasized as a character to distinguish between genera (Buyck and Horak, 1999).

Therefore, it is concluded that all African *Lactarius* species found so far form ectomycorrhizas. A similar conclusion is accepted for *Russula* and most other genera with known ectomycorrhizal members.

Distribution Patterns of *Lactarius* Species in Africa

In North America (Hesler and Smith, 1979) and Africa (this study), a possible preference of certain *Lactarius* species for certain soil types (e.g. acid or calcareous), a character frequently stressed in Europe (e.g. Verbeken *et al.*, 1999), has received less attention.

Phytogeographic regions are also important for the distribution patterns of fungi. In Europe, several *Lactarius* species are restricted to the Mediterranean area (e.g. *Lactarius cistophilus* Bon & Trimbach), the arctic–alpine zone (e.g. *Lactarius dryadophilus* Kühner), the boreal zone (e.g. *Lactarius hysginoides* Korhonen & T. Ulvinen) or to the atlantic–temperate zone (e.g. *Lactarius circellatus* Fr.). Similarly, Hesler and Smith (1979) distinguished four 'mushroom provinces' in North America in which a large number of endemics have evolved. The author's observations for tropical Africa suggest a high phytogeographical and ecological specificity. Although many regions of tropical Africa remain mycologically unexplored (e.g. Sudanian woodlands), it is evident that most *Lactarius* species, also *Russula* (Buyck, 1999) and most other ectomycorrhizal groups, have been collected either in open (miombo, *Uapaca* woodland) or in dense forest types (e.g. rain-, riparian, gallery, swamp and dry evergreen forests) but rarely in both (Table 2.3).

A relatively large number of species (e.g. *Lactarius brunnescens* Verbeken, *Lactarius densifolius* Verbeken & Karhula, *Lactarius edulis* Verbeken & Buyck, *Lactarius heimii*, *Lactarius longisporus* Verbeken, *Lactarius medusae* Verbeken, *Lactarius urens* Verbeken, *Lactarius velutissimus* Verbeken & Buyck) seems to be restricted to the Zambezian region as defined by White (1983). The Sudanian region, although poor in ectotrophs and lacking important mycorrhizal tree genera such as *Brachystegia* and *Julbernardia* (Thoen, 1993), remains to be explored, but the recent discovery in Benin of *Lactarius kabansus* Pegler & Piearce and *Lactarius gymnocarpoides* Verbeken, both common species in the Zambezian area, suggests the existence of a number of species that occur in both Sudanian and Zambezian woodland.

Again the distribution of *L. gymnocarpus* suggests that the Guineo–Congolian region is another centre of endemic mycorrhizal fungi (Fig. 2.1), which could be extended to the East Malagasy region as demonstrated by *Lactarius acutus* R. Heim, *Lactarius melanogalus* R. Heim ex R. Heim, *Lactarius rubroviolascens* R. Heim and *Lactarius annulatoangustifolius* (Beeli) Buyck. Specialized vegetation such as *Uapaca* gallery forest possibly allows Guineo-Congolian ectomycorrhizal fungi to invade the Zambezian region. Apparently,

Table 2.3. Phytogeographical and ecologial specificity of African *Lactarius* species: occurrence of the known species in woodland ecosystems (WL: miombo woodland, Sudanian woodland) and forest ecosystems (RF: rain-, dry evergreen, gallery and swamp forests). +: all collections studied, (+): minority of collections studied.

	WL	RF		WL	RF
acutus		+	*longipes*		+
adhaerens		+	*longisporus*	+	
albocinctus	+		*luteopus*	+	
amarus		+	*medusae*	+	
angustus		+	*melanodermus*		+
annulatoangustifolius	(+)	+	*melanogalus*		+
arsenei		+	*nonpiscis*	+	
atroolivinus	+		*nudus*		+
aurantiifolius	+		*orientalis*	+	
aureifolius	+		*pelliculatus*		+
badius		+	*phlebonemus*	(+)	+
baliophaeus	+	(+)	*phlebophyllus*		+
barbatus	+		*pisciodorus*		+
brachystegiae	+		*pruinatus*	+	
brunnescens	+		*pseudogymnocarpus*		+
caperatus		+	*pseudolignyotus*		+
carmineus	+		*pseudotorminosus*		+
chamaeleontinus	(+)	+	*pseudovolemus*		+
chromospermus	+		*pulchrispermus*	+	
claricolor		+	*pumilus*	+	
congolensis		+	*pusillisporus*	+	
corbula		+	*roseolus*	+	
cyanovirescens	+		*rubiginosus*	+	
denigricans	+		*rubroviolascens*		+
densifolius	+		*rumongensis*	+	(+)
edulis	+		*russuliformis*		+
emergens	+		*ruvubuensis*		+
flammans	+		*saponaceus*	+	
fulgens		+	*sesemotani*	+	(+)
var. *africanus*		+	*striatus*		+
goossensiae		+	*subamarus*	+	
griseogalus	+		*sulcatulus*	+	(+)
gymnocarpoides	+		*sulcatus*	+	
gymnocarpus		+	*tanzanicus*	+	
heimii	+		*tenellus*	+	
hispidulus		+	*undulatus*		+
indusiatus	+		var. *rasilis*		+
inversus		+	*urens*	+	
kabansus	+		*velutissimus*	+	
kalospermus		+	*volemoides*	+	
kivuensis		+	*xerampelinus*	+	
laevigatus	+		*zenkeri*		+
latifolius		+			

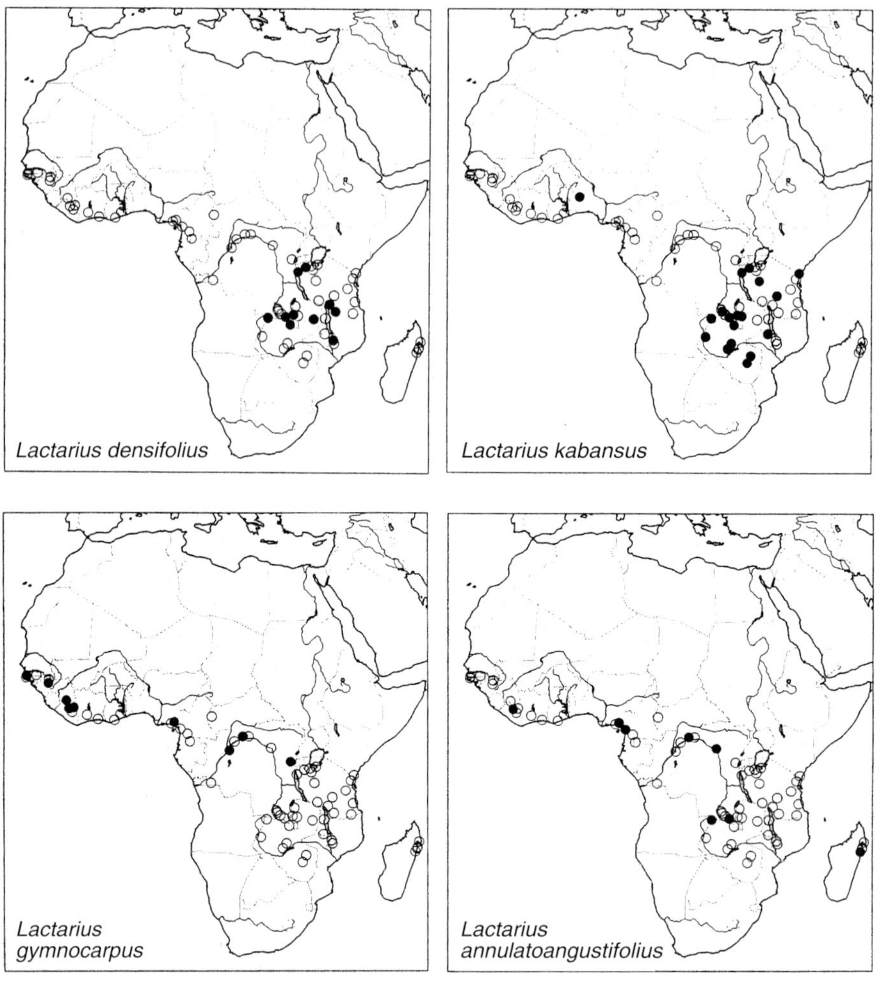

Fig. 2.1. Distribution of *Lactarius densifolius, L. kabansus, L. gymnocarpus* and *L. annulatoangustifolius*; open circle, collection sites; closed circle, taxon present.

the distribution of many African ectomycorrhizal fungi (including Madagascar) is linked to the contrast between open woodlands and dense forest.

Host Specificity

In Europe, with its numerous monospecific or species-poor forests, many *Lactarius* species are associated with only one tree genus (e.g. *Lactarius salmonicolor* R. Heim & Leclair with *Abies, Lactarius quietus* (Fr.: Fr.) Fr. with *Quercus, Lactarius lilacinus* (Lasch: Fr.) Fr. with *Alnus*). Literature data, herbarium notes and the author's own field observations suggest that this host specificity is lacking in Africa. Most *Lactarius* species fruit under several tree genera in African forests, which are usually much richer in tree species than temperate forests (Thoen and Bâ, 1989; Härkönen *et al.*, 1993; Munyanziza and Kuyper, 1995; Buyck *et al.*, 1996), and this more generalist relationship can be extrapolated to most ectomycorrhizal fungi. More observations on monospecific tropical vegetation types might demonstrate the existence of a higher degree of specificity for some species, such as *Lactarius chromospermus* Pegler which has been collected exclusively under *Brachystegia* species at many localities (Buyck and Verbeken, 1995; Verbeken, 1996d; Verbeken *et al.*, 2000).

Phenology

Intensive collecting in the Rumonge region in Burundi made it possible to make some preliminary observations on the fruiting periodicity of *Lactarius* species in miombo woodland. An initial short period of rainfall (end of November–January) is usually separated from a long period of rain from (15 February) March to April (–15 May) by a short dry season (January–February). From the available data over a period of 3 years, it became clear that the fructification of several species is virtually confined to one of these wet periods (Fig. 2.2). In particular, *Lactarius barbatus* Verbeken and *L. densifolius* were collected only during the first rains. *L. kabansus* and *Lactarius subamarus* Verbeken were especially abundant at the beginning of this season, but were sporadically collected during the whole rainy season. In contrast, some species (such as *L. edulis*, *Lactarius rumongensis* Verbeken and *Lactarius pusillisporus* Verbeken) were collected only during the long period of rainfall. This second period yielded a higher number of species, but for a complete inventory in seasonally dry African forest collecting should cover both rainfall periods over several years.

Conclusion

At present, about 400 taxa of putative ectomycorrhizal fungi belonging to *Russulaceae*, *Boletales*, *Cantharellaceae*, *Amanita*, *Cortinariaceae*, various sequestrate fungi (Castellano *et al.*, 2000) and *Scleroderma*, have been described or reported from tropical African forests and woodlands, but it is estimated that at least another 400 taxa remain to be described. The mycorrhizal status seems doubtful for only a restricted number of *Amanita* species and boletes.

Except for *Scleroderma*, a very high proportion of the species are endemic to the studied area (including Madagascar). Judging from an extensive taxonomic study of the genus *Lactarius* (*Russulaceae*), a high degree of host specificity, as known in temperate regions, appears to be absent, but many species are restricted to either woodland vegetation (miombo) or forest vegetation (rain-, dry evergreen and gallery forests). In miombo woodland, many species may

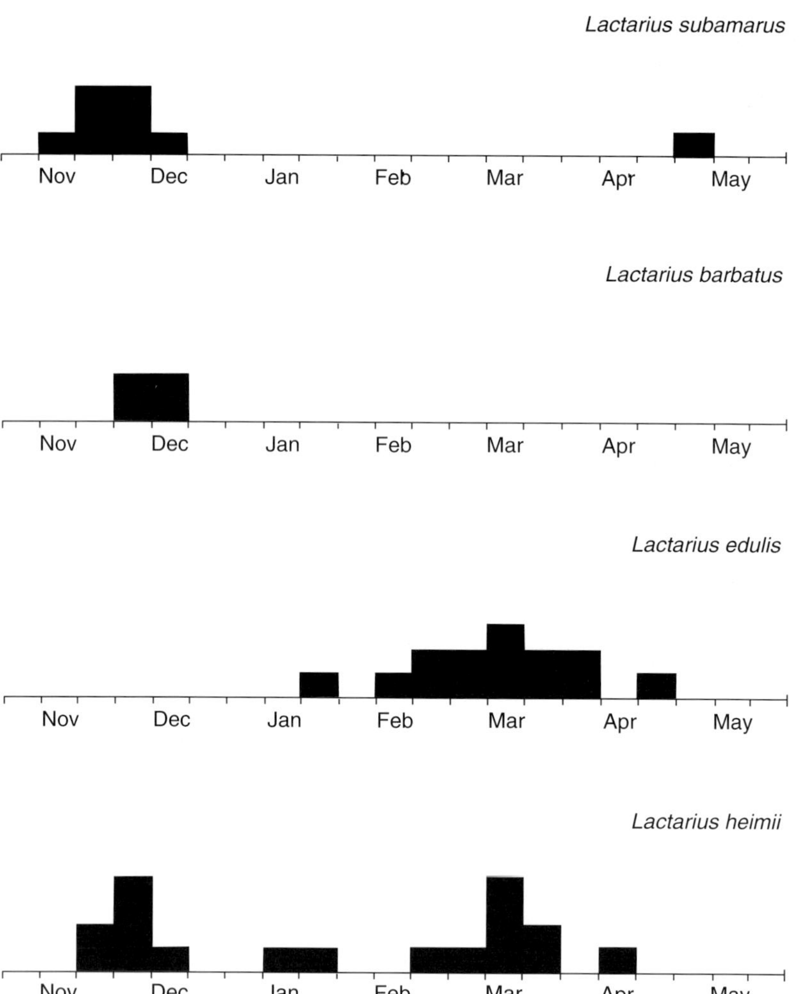

Fig. 2.2. Phenology (based on number of collections over a 3 year period) of *Lactarius subamarus, L. barbatus, L. edulis* and *L. heimii* in the Rumonge district (Burundi).

occur in a restricted area, but some species occur in both the Sudanian and the Zambezian woodland. Several miombo woodland species have a restricted fruiting period and occur during only part of the season.

Acknowledgements

We sincerely thank Benoît Nzigidahera and Cassien Niyondiko for additional collecting and field assistance in Burundi, the Dean of the Faculty of Science and the Head of the Belgian-Burundian cooperation for field facilities, and the Belgian National Fund for Scientific Research (FWO) for sponsoring the expeditions of A. Verbeken. Ruben Walleyn and Roy Watling are acknowledged for suggestions and comments on an earlier version of this chapter.

References

Allsopp, N. and Stock, W.D. (1993) Mycorrhizal status of plants growing in the Cape Floristic Region, South Africa. *Bothalia* 23, 91–104.

Arnolds, E., Kuyper, T.W. and Noordeloos, M.E. (eds) (1995) *Overzicht van de paddestoelen in Nederland.* Nederlandse Mycologische Vereniging, Wijster, The Netherlands.

Baxter, A.P. (1990) A large bolete from South Africa. *Mycologist* 4, 91–93.

Beeli, M. (1928) Contribution à l'étude de la flore mycologique du Congo. Fungi Goossensiani VI. *Bulletin de la Société Royale Botanique de Belgique* 61, 78–107.

Beeli, M. (1935) *Amanita, Volvaria. Flore Iconographique des Champignons du Congo* 1, 1–28.

Bresinsky, A. and Stangl, J. (1970) Beiträge zur Revision M. Britzelmayrs 'Hymenomyceten aus Südbayern'. 10. Die Gattung *Lactarius* in der weiteren Umgebung Augsburgs. *Zeitschrift für Pilzkunde* 36, 41–59.

Buyck, B. (1993) *Russula* I (*Russulaceae*). *Flore Illustrée des Champignons d'Afrique Centrale* 15, 335–408.

Buyck, B. (1994a) *Russula* II (*Russulaceae*). *Flore Illustrée des Champignons d'Afrique Centrale* 16, 411–539.

Buyck, B. (1994b) Ectotrophy in tropical African ecosystems. In: Seyani, J.H. and Chikuni, A.C. (eds) *Proceedings XIIIth Plenary Meeting AETFAT Congress, Malawi, 2–11 April 1991*, Vol. 1. Montfort Press and Papulan Publications, Zambia, pp. 705–718.

Buyck, B. (1997) *Russula* III (*Russulaceae*). *Flore Illustrée des Champignons d'Afrique Centrale* 17, 545–598.

Buyck, B. (1999) Il contributo della micoflora tropicale africana per una classificazione piu naturale delle *Russulaceae. Pagine di Micologia* 12, 53–58.

Buyck, B. and Eyssartier, G. (1999) Two new species of *Inocybe* (*Cortinariaceae*) from African woodland. *Kew Bulletin* 54, 675–681.

Buyck, B. and Horak, E. (1999) New taxa of pleurotoid *Russulaceae. Mycologia* 9, 532–537.

Buyck, B. and Verbeken, A. (1994) Diversity, phenology and host specificity of *Russulaceae* in *Brachystegia* woodland in Burundi (Central Africa). *Abstracts of the Fifth International Mycological Congress, 14–21 August 1994. Vancouver, Canada*, p. 27.

Buyck, B. and Verbeken, A. (1995) Studies in tropical African *Lactarius* species. 2. *Lactarius chromospermus* Pegler. *Mycotaxon* 56, 427–442.

Buyck, B., Thoen, D. and Watling, R. (1996) Ectomycorrhizal fungi of the Guinea–Congo Region. *Proceedings of the Royal Society of Edinburgh* 104B, 313–333.

Castellano, M., Verbeken, A., Walleyn, R. and Thoen, D. (2000) Some new or interesting sequestrate *Basidiomycota* from African woodlands. *Karstenia* 40, 11–21.

Chiu, W.F. (1945) The *Russulaceae* of Yunnan. *Lloydia* 8, 31–59.

Connel, J.H. and Lowman, M.D. (1989) Low-diversity tropical rain forests: some possible mechanisms for their existence. *American Naturalist* 134, 88–119.

Courtecuisse, R. and Duhem, B. (1994) *Les champignons de France.* Eclectis, Paris.

Demoulin, V. and Dring, D.M. (1971) Two new species of *Scleroderma* from tropical Africa. *Transactions of the British Mycological Society* 56, 163–165.

Demoulin, V. and Dring, D.M. (1975) Gasteromycetes of Kivu (Zaire), Rwanda and Burundi. *Bulletin du Jardin Botanique de Belgique* 45, 339–372.

Dennis, R.W.G. (1995) *Fungi of South East England.* Royal Botanic Gardens, Kew, UK.

Dissing, H. and Lange, M. (1962) Gasteromycetes of Congo. *Bulletin du Jardin Botanique de l'Etat* 32, 325–416.

Doidge, E.M. (1950) The South African fungi and lichens to the end of 1945. *Bothalia* 5, 1–1049.

Eyssartier, G. and Buyck, B. (1999a) Contribution à la systématique du genre *Cantharellus* en Afrique tropicale: étude de quelques espèces rouges. *Belgian Journal of Botany* 131, 139–149.

Eyssartier, G. and Buyck, B. (1999b) Contribution à un inventaire mycologique de Madagascar. 2. Nouveaux taxa dans le genre *Cantharellus. Mycotaxon* 70, 203–211.

Fassi, B. and Moser, M. (1991) Mycorrhizae in the natural forest of tropical Africa and the neotropics. In: Fontana, A. (ed.) *Funghi, Piante e Suolo.* Consiglio Nazionale delle Ricerche, Torino, Italy, pp. 157–202.

Guzmán, G., Bandala, V.M., Montoya, L. and Saldarriaga, Y. (1989) Nuevas evidencias sobre las relaciones micoflorísticas entre Africa y el Neotropico. El genero *Rugosospora* Heinem. (Fungi, *Agaricales*). *Brenesia* 32, 107–112.

Härkönen, M., Buyck, B., Saarimäki, T. and Mwasumbi, L. (1993) Tanzanian mushrooms and their uses 1. *Russula. Karstenia* 33, 11–50.

Härkönen, M., Saarimäki, T. and Mwasumbi, L. (1994) Tanzanian mushrooms and their uses 4. Some reddish edible and poisonous *Amanita* species. *Karstenia* 34, 47–60.

Heim, R. (1938) Les Lactario-russulés du domaine oriental de Madagascar. *Prodrome de la Flore Mycologique de Madagascar et Indépendances* 1, 1–196.

Heim, R. (1955) Les lactaires d'Afrique intertropicale (Congo belge et Afrique noire française). *Bulletin du Jardin Botanique de l'Etat* 25, 1–91.

Heinemann, P. (1954) *Boletineae. Flore Iconographique des Champignons du Congo* 3, 49–80.

Heinemann, P. (1958) Champignons récoltés au Congo belge par Madame M. Goossens-Fontana. III. *Cantharellineae. Bulletin du Jardin Botanique de l'Etat* 28, 385–438.

Heinemann P. (1966a) *Cantharellineae* du Katanga. *Bulletin du Jardin Botanique de l'Etat* 36, 335–352.

Heinemann P. (1966b) *Hygrophoraceae, Laccaria* et *Boletineae* II (complément). *Flore Iconographique des Champignons du Congo* 15, 279–308.

Heinemann, P. and Rammeloo, J. (eds) (1972–1997) *Flore Illustrée des Champignons d'Afrique Centrale.* National Botanic Garden of Belgium, Meise, Belgium.

Heinemann, P. and Rammeloo, J. (1980) *Gyrondontaceae* p.p. (*Boletineae*) *Flore Illustrée des Champignons d'Afrique Centrale* 7, 128–131.

Heinemann, P. and Rammeloo, J. (1983) *Gyrodontaceae* (*Boletineae*). *Flore Illustrée des Champignons d'Afrique Centrale* 10, 171–198.

Heinemann, P. and Rammeloo, J. (1986) *Paxillaceae* (*Boletineae*). *Flore Illustrée des Champignons d'Afrique Centrale* 12, 263–271.

Heinemann, P. and Rammeloo, J. (1987) *Phylloporus* (*Boletineae*). *Flore Illustrée des Champignons d'Afrique Centrale* 13, 275–309.

Heinemann, P. and Rammeloo, J. (1989a) *Suillus* (*Boletaceae, Boletineae*). *Flore Illustrée des Champignons d'Afrique Centrale* 14, 313–319.

Heinemann, P. and Rammeloo, J. (1989b) *Tubosaeta* (*Xerocomaceae, Boletineae*). *Flore Illustrée des Champignons d'Afrique Centrale* 14, 321–335.

Hesler, L.R. and Smith, A.H. (1979) North American species of *Lactarius*. University of Michigan Press, Ann Arbor, Michigan.

Högberg, P. (1982) Mycorrhizal associations in some woodland and forest trees and shrubs in Tanzania. *New Phytologist* 92, 407–415.

Högberg, P. and Nylund, J.-E. (1981) Ectomycorrhizae in coastal Miombo woodland of Tanzania. *Plant and Soil* 63, 283–289.

Högberg, P. and Piearce, G.D. (1986) Mycorrhizas in Zambian trees in relation to host taxonomy, vegetation type and successional patterns. *Journal of Ecology* 74, 775–785.

Karhula, P., Härkönen, M., Saarimäki, T., Verbeken, A. and Mwasumbi, L. (1998) Tanzanian mushrooms and their uses 6. *Lactarius. Karstenia* 38, 49–68.

Krieglsteiner, G.J. (1991) *Verbreitungsatlas der Grosspilze Deutschlands (West)*. Band 1: *Ständerpilze*. Teil A: *Nichtblätterpilze*. Ulmer Verlag, Eugen, Germany.

Maghembe, J.A. and Redhead, J.F. (1980) A survey of ectomycorrhizal fungi in pine plantations in Tanzania. *East African Agricultural Forestry Journal* 45, 203–206.

May, T.W. and Wood, A.E. (1997) *Catalogue and Bibliography of Australian Macrofungi*. 1 *Basidiomycota p.p. Fungi of Australia* 2A. Australian Biological Resources Study, Canberra.

Munyanziza, E. and Kuyper, T.W. (1995) Ectomycorrhizal synthesis on seedlings of *Afzelia quanzensis* Welw. using various types of inoculum. *Mycorrhiza* 5, 283–287.

Pearson, A.A. (1950) Cape agarics and boleti. *Transactions of the British Mycological Society* 33, 276–316.

Pegler, D.N. and Fiard, J.P. (1979) Taxonomy and ecology of *Lactarius* (Agaricales) in the Lesser Antilles. *Kew Bulletin* 33, 601–628.

Pegler, D.N. and Rayner, R.W. (1969) A contribution to the agaric flora of Kenya. *Kew Bulletin* 23, 347–412.

Pegler, D.N. and Shah-Smith, D. (1997) The genus *Amanita* (*Amanitaceae, Agaricales*) in Zambia. *Mycotaxon* 61, 389–417.

Peyronel, B. and Fassi, B. (1957) Micorrize ectotrofiche in una Cesalpiniacea del Congo Belga. *Atti della accademia delle scienze di Torino* 91, 569–576.

Reid, D.A. and Eicker, A. (1991) South African fungi: the genus *Amanita*. *Mycological Research* 95, 80–95.

Robyns, W. (ed.) (1935–1972) *Flore Iconographique des Champignons du Congo*. National Botanic Garden of Belgium, Meise, Belgium.

Singer, R. (1986) *The Agaricales in Modern Taxonomy*, 4th edn. Koeltz, Koenigstein, Germany.

Singer, R. and Araujo, I.A. (1986) Litter decomposition and ectomycorrhizal basidiomycetes in an igapo forest. *Plant Systematics and Evolution* 153, 107–117.

Taylor, F.W., Thamage, D.M., Roth-Bejerano, N. and Kagan-Zur, V. (1995) Notes on the Kalahari desert truffle, *Terfezia pfeilii*. *Mycological Research* 99, 874–878.

Thoen, D. (1993) Looking for ectomycorrhizal trees and ectomycorrhizal fungi in tropical Africa. In: Isaac, S., Frankland, J.C., Watling, R. and Whalley, A.J.S. (eds) *Aspects of Tropical Mycology*. Cambridge University Press, Cambridge, pp. 193–205.

Thoen, D. and Bâ, A.M. (1989) Ectomycorrhizas and putative ectomycorrhizal fungi of *Afzelia africana* Sm. and *Uapaca guineensis* Müll. Arg. in southern Senegal. *New Phytologist* 113, 549–559.

Thoen, D. and Ducousso, M. (1989) Champignons et ectomycorrhizes du Fouta Djalon. *Bois et Forêts des Tropiques* 221, 45–63.

Van der Westhuizen, G.C.A. and Eicker, A. (1987) Some fungal symbionts of ectotrophic mycorrhizae of pines in South Africa. *South African Forestry Journal* 143, 20–24.

Van der Westhuizen, G.C.A. and Eicker, A. (1994) *Field Guide to the Mushrooms of Southern Africa.* Struik Publishers, Cape Town, South Africa.

Verbeken, A. (1995) Studies in tropical African *Lactarius* species. 1. *Lactarius gymnocarpus* R. Heim ex Singer and allied species. *Mycotaxon* 55, 515–542.

Verbeken, A. (1996a) Studies in tropical African *Lactarius* species. 4. Species described by P. Hennings and M. Beeli. *Edinburgh Journal of Botany* 52, 49–79.

Verbeken, A. (1996b) Studies in tropical African *Lactarius* species. 3. *Lactarius melanogalus* and related species. *Persoonia* 16, 209–223.

Verbeken, A. (1996c) New taxa of *Lactarius* (*Russulaceae*) in tropical Africa. *Bulletin du Jardin Botanique National de Belgique* 65, 197–213.

Verbeken, A. (1996d) Biodiversity of the genus *Lactarius* Pers. in tropical Africa. Part 1 and 2. Doctoral thesis, Ghent University, Belgium.

Verbeken, A. (1998a) Studies in tropical African *Lactarius* species. 5. A synopsis of the subgenus *Lactifluus* (Burl.) Hesler & A.H. Sm. emend. *Mycotaxon* 66, 363–386.

Verbeken, A. (1998b) Studies in tropical African *Lactarius* species. 6. A synopsis of the subgenus *Lactariopsis* (Henn.) R. Heim emend. *Mycotaxon* 66, 387–418.

Verbeken, A. (2000) Studies in tropical African *Lactarius* species. 8. A synopsis of the subgenus *Plinthogali*. *Persoonia* 17, 377–406.

Verbeken, A. and Walleyn, R. (2000) Studies in tropical African *Lactarius* species. 7. A Synopsis of the section *Edules* and a review on the edible species. *Belgian Journal of Botany* 66, 175–184.

Verbeken, A., Fraiture, A. and Walleyn, R. (1999) Observations on the genus *Lactarius* in Belgium, with a special reference to its section *Plinthogali*. *Belgian Journal of Botany* 131, 211–222.

Verbeken, A., Walleyn, R., Sharp, C. and Buyck, B. (2000) Studies in tropical African *Lactarius* species. 9. Records from Zimbabwe. *Systematics and Geography of Plants* 70, 181–215.

Walleyn, R. and Verbeken, A. (1999) Notes on the genus *Amanita* in sub-saharan Africa. *Belgian Journal of Botany* 66, 175–184.

Watling, R. (1992) *Fungus Flora of Shetland.* Royal Botanic Garden, Edinburgh, UK.

Watling, R. (1993) Comparison of the macromycete biotas in selected tropical areas of Africa and Australia. In: Isaac, S., Frankland, J.C., Watling, R. and Whalley, A.J.S. (eds) *Aspects of Tropical Mycology*. Cambridge University Press, Cambridge, pp. 171–191.

Watling, R. and Turnbull, E. (1993) Boletes from South and East Central Africa I. *Edinburgh Journal of Botany* 49, 343–361.

Watling, R. and Turnbull, E. (1994) Boletes from South and East Central Africa II. *Edinburgh Journal of Botany* 51, 331–353.

White, F. (1983) *The Vegetation of Africa. A Descriptive Memoir to Accompany the Unesco/AETFAT/UNSO Vegetation Map of Africa.* Unesco, Paris.

The Occurrence and Distribution of Putative Ectomycorrhizal Basidiomycetes in a Regenerating South-east Asian Rainforest

3

R. WATLING,[1,2] S.S. LEE[3] AND E. TURNBULL[2]

[1]*Caledonian Mycological Enterprises, 26 Blinkbonny Avenue, Edinburgh EH4 3HU, UK;* [2]*Royal Botanic Garden, Edinburgh EH3 5LR, UK;* [3]*Forest Research Institute Malaysia, Kepong, 52109 Kuala Lumpur, Malaysia*

Introduction

The forests of Malaysia and large parts of South-east Asia are dominated by the ectomycorrhizal *Dipterocarpaceae*, one of the most important timber trees in the region. In 1995, as part of the Malaysia–United Kingdom Programme on Conservation, Management and Development of Forest Resources, a study was carried out in Pasoh Forest Reserve to enumerate the occurrence of putative ectomycorrhizal fungi in logged and unlogged forests (Watling *et al.*, 1998). While part of the results of that study has been reported elsewhere (Watling *et al.*, 1998; Lee *et al.*, 2001), a comprehensive discussion of the results obtained from the logged over forest is presented here.

The 2450 ha Pasoh Forest Reserve, managed by the Forest Research Institute Malaysia (FRIM), is located approximately 140 km south-east of the capital city, Kuala Lumpur, in the state of Negri Sembilan, and consists of lowland dipterocarp forest of the Keruing-Meranti type. The forest here is floristically very rich; a total of 335,256 stems of at least 1 cm diameter at breast height (dbh) and above, belonging to 814 species, 294 genera and 78 families, has been recorded within an area of 50 ha. The most common plant families are the *Euphorbiaceae* and *Annonaceae* among the smaller trees, and the *Dipterocarpaceae*, *Leguminosae* and *Burseraceae* among the larger trees. For trees above 30 cm dbh, the most abundant species is *Shorea leprosula* Miq., a member of the *Dipterocarpaceae* (Lee, 1995). The core of the Reserve is pristine in condition and is surrounded by a zone of regenerating lowland forest buffering it from oil palm plantations and other mixed agricultural land. Thus logged and unlogged forests are adjacent to one another in the same forest type. The significance of this was recognized in the late 1960s and from 1970 to 1978 Pasoh was the site of intensive research on lowland rainforest ecology and dynamics. In continuation of this long-term study in early 1995, five 100 m^2 plots were established in Compartment 25 of this forest reserve as part of a study on forest regeneration. Compartment 25 was selectively logged in 1955, under the Malayan Uniform System (MUS) and is now considered to be a regenerating forest (Manokaran, 1998).

Methods

As part of a general mycological survey of both the logged and unlogged areas under the Malaysia–United Kingdom Programme, five 1 ha plots (A–E) were visited twice a year between 1995 and 1997, once in February–March (spring) and again in August–September (autumn). These visits coincided with the fungal fruiting seasons that occur at the end of a prolonged dry spell. Corner (1935) observed that there are two general fruiting seasons in the Malay peninsula, the first in March and the second about August or September. The study plots were in two parallel blocks separated by a corridor 100 m wide and each plot was separated from its neighbours by 100 m, so forming a 300 m × 500 m framework, although, unfortunately, Plot F was never delimited.

Each plot was surveyed both in the spring and in the autumn twice with no less than three days between each visit. The possibility of recording the same basidiome twice, as seen in studies in temperate woodland (Richardson, 1970), was not considered a limitation because of the short life span of the basidiomes caused by high rainfall and humidity encouraging decay and attack by hyperparasites and animal mycophagy.

The recording of ectomycorrhizal fungi simply by the occurrence of their basidiomes is fraught with danger as the mycelium of the fungus may travel several metres before fruiting, and therefore its basidiomes may appear under the canopy of a totally different tree. Observations in boreal communities show that ectomycorrhizal fungal associates generally fruit towards the edge of the canopy of the host and careful soil excavation shows the physical connection between fungus and tree. Fruiting position therefore has been used as the criterion applied in the present study, although the limitations of this approach have been discussed by Watling (1995). Limited soil excavations have been carried out in lowland rainforest (Lee *et al.*, 2001), but the direct link between host and fungus was not determined in the present survey due to the time-consuming nature of such work, nor was any concept of an individual fungus considered.

Each 1 ha plot was divided into 10 m × 10 m squares, and three persons walking abreast, approximately 3 m apart, covered each row of ten squares in turn, from 1–10, 11–20, etc. All basidiomes present were noted and a tentative identification, often possible only to genus, was made. The identity of the tree(s) nearest to the basidiomes was also recorded. In these plots all trees above 10 cm dbh were tagged, mapped and identified. Fungi whose basidiomes were found beneath the canopy of a single tree were assumed to be putative ectomycorrhizal partners of that tree. In cases where canopies overlapped, then each tree was noted, but rarely was it necessary to record more than two putative tree associates per basidiome. The tree number provided positional data within the plot.

Voucher materials were collected from the first two examples of each morphotype encountered. However, material was always collected if there was any uncertainty about morphotype matching. All the material was carefully collected as outlined by Henderson *et al.* (1969), labelled and processed for identification and herbarium retention. Material has been deposited in the Royal Botanic Garden, Edinburgh (E) with duplicates when possible at FRIM (KEP). When the same fungus was again encountered, additional material was also kept but thereafter the morphotype was simply noted; if there was any uncertainty, then all collections were retained. Only one representative collection was made if the fungus was already known from the main survey area which had been studied since 1992; for example *Craterellus cornucopioides* (L.: Fr.) Pers. (Fig. 3.1).

Fig. 3.1. *Craterellus cornucopioides* (var. *microsporus* Corner); September 1996.

Results and Discussion

Three hundred and twenty-four distinct collections of basidiomes were recorded over 2.5 years (1995–1997) from five plots (A–E) and adjacent areas (Table 3.1). The records were distributed among 14 families of basidiomycetes and were mostly of the *Russulaceae*, especially members of the genus *Russula* itself. This family not only had the greatest numbers of basidiomes (218) but also the greatest number of different species (morphotypes) recorded in this study. Members of the *Amanitaceae* and *Boletaceae* were ranked a distant second and third with 26 and 22 collections of basidiomes respectively. In the surrounding unlogged areas, studies conducted over a 6 year period (Lee *et al.*, 2002) indicated that the number of species of members of families such as the *Boletaceae* and *Amanitaceae* approached more closely the numbers of the *Russulaceae*, although the last family generally produced more basidiomes per colony. In the present study, most records were made in 1996 when 114 records were obtained in spring and 82 in autumn (Table 3.1). Over the duration of the study, Plot A yielded the highest number of collections of basidiomes (74) followed by Plots C (68) and E (53), while Plots B and D were much poorer with only 30 and 21 collections respectively. Basidiomes recorded from Plot C in spring 1996 were mostly of *Russula crustosa* Peck and *R.* Schaeff.: Fr.

Although the number of basidiomes collected on one day in a single plot provided an indication of fruiting abundance, it did not necessarily reflect the diversity of putative ectomycorrhizal species present. In contrast to the results above, Plots A and E had the highest species diversity with 41 and 38 different species respectively while Plot D had the lowest number of species with 14 only (Table 3.2). Plots B and C had 23 and 26 species of putative ectomycorrhizal fungi respectively.

The 324 collections of basidiomes recorded in this study were assigned to 95 different taxa of which 52 were identified and named to at least species level. Ninety-two taxa were found within the five 1 ha plots and an additional three species in the areas just outside the set plots (Table 3.2).

The fruiting occurrence of several species of fungi appeared to be quite localized; for example *Astrosporina* (*Inocybe*) *angustifolia* Corner & Horak, *Inocybe sphaerospora* Kobayasi (Fig. 3.2) and *Scleroderma sinnamariense* Mont. were found only in Plot A and *Horakiella* sp. (Fig. 3.3) only in Plot E (Table 3.2). Members of some families were found on every visit, for example many species of *Russulaceae*, and *C. cornucopioides*. However, other species/families were much more sporadic in fruiting and some, such as *Hydnum repandum* L.: Fr. (Fig. 3.4) and *Pisolithus aurantioscabrosus* Watling (Fig. 3.5) which had been prominent in spring 1995 and autumn 1996 respectively, were not fruiting on other visits (Table 3.2). Members of the *Amanitaceae* were prominent in the plots during autumn 1996 and spring 1997 but were not encountered in 1995 (Tables 3.1, 3.2). They were, however, fruiting elsewhere outside the plots during that time. Two hypogeous species were found in the plots: *Dendrogaster cambodgensis* Pat.

Table 3.1. Number of collections of basidiomes of fungi recorded from within and adjacent to (>) five 1 ha plots (A–E) at Pasoh Forest Reserve, 1995–1997.

Family	>A	A	A>B	B	B>C	C	D	D>E	E	A>D	B>E	Total
Spring 1995												31
Amanitaceae	1											1
Boletaceae	4	3										7
Cantharellaceae	1	1										2
Cortinariaceae		3										3
Russulaceae	3	5	3			1	1		1	1		15
Secotiaceae		1										1
Tricholomataceae		1		1								2
Autumn 1995												36
Cantharellaceae	2			1			1	1				5
Entolomataceae	1											1
Hydnaceae		2	1								1	4
Hymenogastraceae		1										1
Russulaceae	7	2	2	1		6			4		2	24
Sclerodermataceae	1											1
Spring 1996												114
Amanitaceae		1							2			3
Boletaceae	1						1					2
Cantharellaceae				1			2		1	1		5
Hydnaceae	1										1	2
Hymenogastraceae											1	1
Russulaceae		23		4		37	6	18	8	1	2	99
Sclerodermataceae									1			1
Secotiaceae						1						1
Autumn 1996												82
Amanitaceae		3		6		1			6			16
Boletaceae		2		3		3			1			9
Cantharellaceae		1		1		1			1			4
Entolomataceae		1		1					2			4
Hymenogastraceae					1							1
Paxillaceae					1							1
Russulaceae		14	1	8		2	2		15			42
Sclerodermataceae				1					1			2
Secotiaceae									1			1
Pisolithaceae				1					1			2
Spring 1997												61
Amanitaceae	1		1		1	2	1					6
Boletaceae		1				1	1		1			4
Cantharellaceae		1			1	2						4
Clavulinaceae						1						1
Cortinariaceae		1										1
Entolomataceae						2						2
Hymenogastraceae				1	1		1					3
Paxillaceae							1					1
Russulaceae	1	6	5		4	8	4		7	3		38
Sclerodermataceae		1										1
Total	24	74	13	30	9	68	21	19	53	6	7	324

Table 3.2. Taxa of putative ectomycorrhizal fungi found within and immediately outside five 1 ha plots (A–E) in Pasoh Forest Reserve, 1995–1997.

	Collecting period				
	1995		1996		1997
Fungus	Spring	Autumn	Spring	Autumn	Spring
Amanitaceae					
Amanita alauda[a]				B	C
A. centunculus[a]					C
A. hemibapha subsp. *similis*[a]				E	
A. obsita[a]				B	
A. perpasta					D
A. pilosella[a]			E		
A. privigna[a]				B	
A. cf. *privigna*[a]				A	
A. tjibodensis[a]				E[b]	
A. tristis[a]				B	
A. vestita[a]				A	
A. virginea[a]			E		
Amanita sp. 2 Corner & Bas			A	E	
Boletaceae					
Boletus peltatus[a]				B	
Boletus sect. *Subpruinosae*					D
Boletus sp.				A	
Pulveroboletus frians[a]				A, E	A
Pulveroboletus nov. sp.[a]					C
Rubinoboletus ballouii[a]	A				
R. ballouii var. *fuscatus*[a]				C	
Tylopilus spinifer comb. prov.					E
T. tristis comb. prov.[a]	A				
Tylopilus sp.				A	
Boletus sp. (xerocomoid)				C	
Cantharellaceae					
Cantharellus ianthinus[a]				A, B, C	
Cantharellus omphalinoides					C, C[b]
Cantharellus sp. B					C
Cantharellus sp.			E		
Craterellus cornucopioides[a]	A	B, D	B	E	A
Clavulinaceae					
Clavulina cartilaginea					C
Cortinariaceae					
Inocybe angustifolia[a]	A				
I. sphaerospora[a]	A				A
Entolomataceae					
Entoloma cf. *burkilliae*				B, E	
E. corneri[a]					C
E. flavidum[a]				E	

Table 3.2. (*Continued*).

	Collecting period				
	1995		1996		1997
Fungus	Spring	Autumn	Spring	Autumn	Spring
E. pallidoflavum				A	
E. pingue					C
Hydnaceae					
Hydnum repandum[a]		A			
Hymenogastraceae					
Dendrogaster cambodgensis[a] (=*Mycoamaranthus*)		A			B, D
Paxillaceae					
Phylloporus orientalis var. *brevisporus*[a]					D
P. rufescens				B[b]	
Pisolithaceae					
Pisolithus aurantioscabrosus[a]				B, E	
Russulaceae					
Lactarius ruginosus[a]			E		
L. ? sumsteinii[a]			E		
Lactarius sect. *Plinthogali #1*[a]		A			C
Lactarius sect. *Plinthogali #2*		E	A		
Lactarius sect. *Russulares*[a]				E	
Lactarius sect. *Piperites*[a]				E	
Russula alboareolata[a]	A, D, E,		A, E		
R. aff. *cutefracta*[a]			E		
R. castanopsidis sect. *Ilicinae?*[a]			A, D		
R. crustosa[a]			C		
R. cyanoxantha[a]				E	D
R. cf. *illota*[a]		B, C		C	
R. japonica[a]	A			A	A, C, E
R. singaporensis[a]			A, D, E	A, D, E	A, D
R. subnigricans[a]			D		A, D
R. cf. *variata*[a]			A, B	A, B, E	A, C, E
R. cf. *vesca*[a]					E
R. violeipes[a]		A	A	A, B	
R. virescens[a]			C, D, E		
Russula ? Echinosporinae[a]			A	A	
Russula Felleinae/Citrinae #1[a]				E	
Russula Felleinae/Citrinae #2				A, B	
Russula Foetentinae #1[a]	C		B, C, D	B, C, D	
Russula Foetentinae #5[a] (*consobrina*-like)				A, E	
Russula Heterophyllae #1[a]				A	
Russula Indolentae #1[a]					E

Continued

Table 3.2. (*Continued*).

	Collecting period				
	1995		1996		1997
Fungus	Spring	Autumn	Spring	Autumn	Spring
Russula Messapicae[a]			A	A	A
Russula Nigricantae #3[a]			C, D	E	
Russula Nigricantae #5				C	
Russula Plorantinae #3[a]					A
Russula ? *Plorantinae* #4				E	
?*Russula 'Minutae'* #3[a]				A, E	
Russula 'Minutae' #4				B	
Field morphotypes					
Russula resembling *Lactarius controversus*[a]				B	
Russula with rich blue purple pileus as in *R. unicolor*					C
Russula resembling *R. nitida*				A	
Russula resembling *R. carminea*[a]				A, E	C
Russula blood-red pileus #2 as in *Emeticinae*			A		A, E, D
Russula rose-red pileus[a]					C
Russula pale pinkish pileus[a]				B	
Russula pinkish violaceous pileus					A
Russula pale pinkish buff pileus, small				A, B	
Russula livid vinaceous pileus			E		
Russula vinaceous brown pileus			A		
Russula brown pileus			A, C, E	C	A, C, E
Russula sepia grey/horn-colour pileus				B, E	C
Russula orange strongly pectinate pileus				B	
Russula white pileus, fragile as in *R. wieneri*				E	
Sclerodermataceae					
Horakiella sp.[a]				E	
Scleroderma sinnamariense[a]					A
Scleroderma sp.				B	
Secotiaceae[a]	A		C	E	
Tricholomataceae					
Tricholoma monsfraseri[a]	A, B				

[a]The species has been found elsewhere at Pasoh either in the logged and/or unlogged forest.
[b]Basidiomes were found no more than 50 cm outside plot and within the canopy of a tree within the plot.

Fig. 3.2. *Inocybe sphaerospora*; March 1996.

(Fig. 3.6) in Plots A, B and D, and a yet to be determined sequestrate species (*Secotiaceae*) in Plots, A, C and E. Species diversity was relatively low in Plots B and D during the periods when they were visited, except in autumn 1996, indicating that a meaningful idea of the diversity of a site cannot be gained in a short time. The number of species of putative mycorrhizal fungi recorded on the

Fig. 3.3. *Horakiella* sp. (top left); *Chamonixia mucosa* Corner and Hawker, not found in plots (bottom left) and *Scleroderma echinatum*; September 1996.

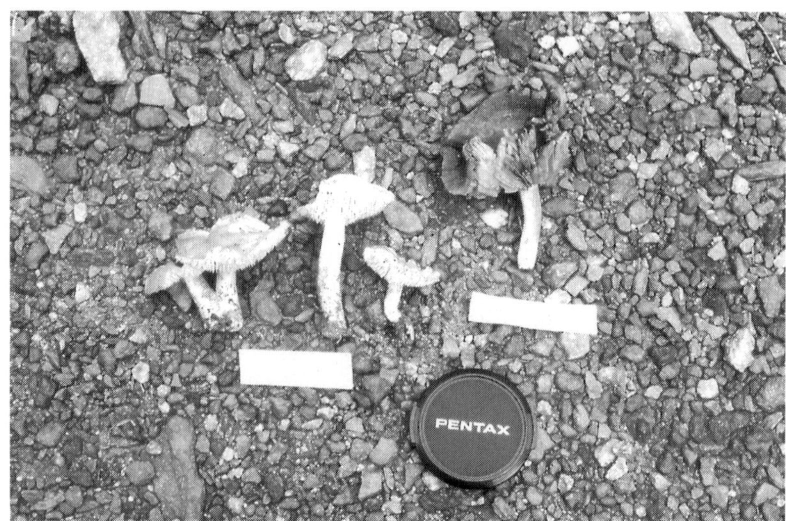

Fig. 3.4. *Hydnum repandum* (left) and blueing *Lactarius* sp. (Sect. *Plinthogali*); September 1995.

plots represented about 32% of the total number for Pasoh *in toto* over a 6-year period (Lee *et al.*, 2002). Overall, the plots were quite poor in fruiting fungal species compared with some adjacent areas of the forest.

The basidiomes from within each plot on the different visits, located by an examination of the tree numbers were not, except for Plot A, evenly spaced throughout the plot and 'hot spots' of fruiting were detected. This may be indirectly linked to phanerogamic competition because of differences in the soil and moisture patterns, which have been shown to be factors in the distribution of palms and associated plants in a 50 ha plot close by (Noor *et al.*, 1998). Thirty-five families of phanerogams, encompassing 104 species, have been found to be associated with basidiomes of putative ectomycorrhizal fungi in the logged-over forest.

Rarely were basidiomes found associated with the same trees in the different years. Thus, in autumn 1996, basidiomes were putatively associated with 28 trees in Plot E, of which only two had previously been found to be associated with basidiomes. Similarly, in Plot A, basidiomes were associated with 27 trees, of which only one had previously been recorded. For the purpose of this study, it was only necessary to recognize and carefully segregate morphotypes. Of the 95 taxa found in the plots during the present study, some are very familiar and known from Japan and temperate countries in Europe and North America; for example *H. repandum* and *C. cornucopioides*. Undoubtedly a number of species were new to science, especially in the *Russulaceae*, and will hopefully receive formal recognition at a later date. Over the 2.5-year duration of the study, no fungus species was found in all five plots. *C. cornucopioides*, however, was found in

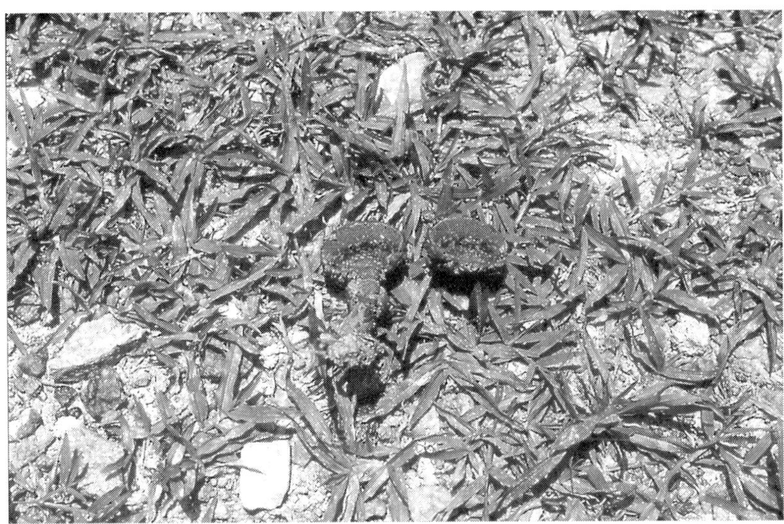

Fig. 3.5. *Pisolithus aurantioscabrosus;* August 1996.

four plots (A, B, D and E) at various times throughout the period of the study (Table 3.2).

Overall, Plot A generally had the highest number of fruiting fungal species during every visit. Some of the more widespread taxa in the other logged areas of the forest reserve were absent from the regenerated forest areas in the plots;

Fig. 3.6. *Dendrogaster cambodgensis* (left) and *Tricholoma rhizophora;* September 1995. The latter was just outside Plot A.

for example *Scleroderma echinatum* (Petri) Guzmán (Fig. 3.3), and several species of boletes and *Amanitaceae*. Surprisingly, species of *Laccaria*, although abundant elsewhere in Pasoh, were not found in the plots.

The same basidiomycete species were often seen fruiting during both the spring and autumn forays, for example several species of *Russula*, and often with similar numbers of basidiomes, complementing Corner's observations made in Singapore more than 70 years ago (Corner, 1940). In contrast, fruiting periodicity differs markedly within temperate woodlands, where although there is a spring fruiting, the main fruiting season is definitely in autumn. This has been attributed to changes in the carbon–nitrogen ratios in the soil (Grainger, 1957) which favour a small and different spring flush. The surge of fruiting which coincided with the autumn visit of 1996 increased the total number of species recorded in the plots by approximately 50%. Such surges may occur regularly, but if the mycologist is not there at the right time, the phenomenon goes unnoticed. The autumn 1996 surge was, however, small when compared with that for 1992 in the main Pasoh Forest Reserve (Lee *et al.*, 2002) where collections of basidiomes made over a few hours on a single day far exceeded the number of basidomes made over several days on other occasions.

Chipp (1921) recognized only three *Russulaceae* from Peninsular Malaysia in his check list of fungi, but our observations, based on the plot information and records from the surrounding areas, indicate at a minimum a 20-fold increase. Corner's work resulted in about a 70% increase in species new to science from Malaysia in the groups that he studied (Hawksworth, 1993). The field observations discussed here confirmed the impression gained by scanning Corner's field notes and illustrations, currently housed in Edinburgh, that there were also many new species of *Russula*. Some basidiomes collected could be placed in known sections of the genus, for example *Foetintinae*, *Plorantinae*, and some even agree with European taxa, for example *Russula violeipes* Quél., a species known also from North America and Japan. Others were not so easily placed, especially a fairly widespread species with long, flexuous, thick-walled, almost cystidioid hyphae clothing the pileipellis. In the genus *Lactarius*, at least three distinct species with blueing milk were recognized, all assignable to the Section *Plinthogali* (Fig. 3.4).

The list of *Entolomataceae* reported here does not include all the collections as there is little or no evidence that species of *Clitopilus*, *Claudopus*, *Leptonia* and *Nolanea* are ectomycorrhizal. However, many members of the genus *Entoloma* are probably host specific and some evidence has been put forward that they form sheathing mycorrhizas in boreal communities (Trappe, 1962) and possibly in South-east Asia (Louis and Scott, 1987). Thus, *Entoloma pingue* Corner & Horak, which was taken as a species of *Tricholoma* in the field, is probably associated with members of the *Dipterocarpaceae*. *Tricholoma mons-fraseri* Corner, not previously recorded since its description (Corner, 1994), is assumed to be mycorrhizal, whereas the relationships of *Tricholoma rhizophora* Corner (Fig. 3.6), found just outside Plot A, and *Tricholoma termitomycetoides* Corner, found in the unlogged forest in the vicinity, might pose some doubt without supporting evidence from soil excavations around the neighbouring roots (Turnbull and Watling, 1999).

Although the five plots were established close to each other in a regenerating forest logged in 1955, the microclimatic conditions found therein were probably quite different, resulting in visible differences in the vegetation and tree distribution. The canopy in Plot A was relatively open with several extremely large stumps, left over from former logging activities, still present. Plot B was located on a slope with wind-fallen trees occupying over one-third of the plot which hampered sighting, locating and collecting of basidiomes. Plot C was relatively level, but the presence of a large windfall gap exposed one side of the plot to very high insolation and ground debris runoff. Plot D was heavily colonized by lianas resulting in a relatively close canopy and large tangles of vegetation. Old logging tracks running diagonally through Plots A and B and wild boar activity in both plots probably led to further changes in the microclimate of these plots. Such factors could have had an influence on the fungal fruiting periodicity and abundance.

An earlier study in Pasoh demonstrated a similar magnitude of putative ectomycorrhizal species diversity between the regenerating and unlogged forest (Watling *et al.*, 1998). This was probably due to the low intensity of logging carried out in the 1950s when many small size dipterocarp trees were left. Indeed, there is evidence that there may even have been an increase in the number of individuals of potential ectomycorrhizal host species in the regenerating forest from the application of the Malayan Uniform System (MUS) silvicultural technique (Manokaran, 1998). It is well known that fungi seek out particular sites in which to fruit and the bare areas between buttress roots, of both living and dead trees, were particularly rich collecting grounds.

Roots of a number of species of dipterocarps have been examined and found to form sheathing mycorrhizas (Lee *et al.*, 1997; Lee, 1998), and some members of the *Leguminosae* are known to be ectomycorrhizal in West and Central Africa (Alexander and Högberg, 1986) and Malaysia (Ahmad, 1989). It may be coincidence that associations have been found with families such as the *Guttiferae*, *Lauraceae* and *Burseraceae*. However, the existence of an ectomycorrhizal association may be highly likely where putative ectomycorrhizal fungi have been consistently collected in association with some species, such as *Diospyros apiculata* Hiern., *Monocarpia marginalis* (Scheff.) Sinclair, *Pentaspadon motleyi* Hook. f. and *Xerospermum noronhianum* Bl.

Although it is well known that members of the *Fagaceae* and *Dipterocarpaceae* form ectomycorrhizal associations (Singh, 1966; Harley and Smith, 1983), many of the families listed came as a surprise and fully enforced our observations in the main forest areas outside the plots; for example *Heritiera simplicifolia* (Mast.) Kosterm. (*Sterculiaceae*) which has been found associated with species of *Horakiella*, *Amanita* and boletes. The genus *Eugenia* (*Myrtaceae*), long suggested as being ectomycorrhizal, has also been found with a range of putative ectomycorrhizal fungi (Table 3.3). The strong association of basidiomes

Table 3.3. Number of collections of basidiomes of putative ectomycorrhizal fungi and taxa of nearby trees within five 1 ha plots in Pasoh Forest Reserve, 1995–1997.

Family	Species	1995		1996			
		Spring	Autumn	Spring	Autumn	1997	Total
Alangiaceae	*Alangium ebenaceum*			1			1
Annonaceae	*Alphonsea maingayi*				1		1
	Monocarpia marginalis			3	2, 2x		5, 2x
	Oncodostigma monosperma		1				1
	Polyalthia jenkinsii		1				1
	Polyalthia rumphii			2			2
	Xylopia ferruginea var. *ferruginea*			1			1
Anacardiaceae	*Buchanania sessifolia*					1	1
	Mangifera quadrifida			1			1
	Parishia insignis			1x			1x
	Pentaspadon motleyi	1		2	2x		3, 2x
Apocynaceae	*Dyera costulata*	1		1			2
Burseraceae	*Canarium littorale*			2	1x	1x	2, 2x
	C. pilosum			1			1
	Dacryodes costata				1x		1x
	D. rostrata			1	1		2
	D. laxa		1				1
	Santiria laevigata		1		1, 2x		2, 2x
	S. griffithii		1		1, 2x		3, 2x
	S. tomentosa		1		1		2
	Triomma malaccensis					1	1
Meliaceae	*Aglaia aspera*					1x	1x
Ebenaceae	*Diospyros apiculata*			1	3		4
	D. sumatrana			1			1
	Diospyros sp. *1*				1, 1x		1, 1x
Euphorbiaceae	*Blumeodendron subrotundifolium*			1x			1x
	Cleistanthus sumatranus			1			1
	Croton laevifolius				1, 1x		1, 1x
	Drypetes kikir				1		1
	D. laevis			1, 1x			1, 1x
	D. pendula				1		1
	Elateriospermum tapos					1	1
	Endospermum malaccense			2	1		3
	Fahrenheitia pendula				1		1
	Koilodepas longifolium				1		1
	Mallotus griffithianus			1x	1		1, 1x
	Neoscortechinia kingii				1, 1x		1, 1x
	N. nicobarica				1		1
	Phyllanthus emblica			1			1
	Pimelodendron griffithianum			5			5
Flacourtiaceae	*Homalium longifolium*				1		1
	Paropsia vareciformis	3, 3x					3, 3x
	Ryparosa kunstleri	1		1		1	3
Guttiferae	*Calophyllum wallichianum*				1, 1x		1, 1x
	Garcinia parvifolia			1x			1x

Table 3.3. (*Continued*).

Family	Species	1995		1996			
		Spring	Autumn	Spring	Autumn	1997	Total
	Mesua ferrea			1	1x		1, 1x
Irvingiaceae	*Irvingia malayana*				1		1
Lauraceae	*Beilschmiedia dictyoneura*			1			1
	Litsea costalus			1			1
Lecythidaceae	*Barringtonia pendula*			1x			1x
Leguminosae	*Archidendron bulbalinum*			5		1x	5, 1x
	Cynometra malaccensis			2		1x	2, 1x
	Intsia palembanica				1x	1	1x
	Koompassia malaccensis			2		1	3
	Millettia atropurpurea					1x	1x
	Parkia speciosa	1		2	2x	1	3, 2x
	Saraca declinata			1			1
	Sindora coriacea				1x		1x
	S. velutina			1			1
Ixonanthaceae	*Ixonanthes icosandra*					3	3
Melastomataceae	*Pternandra echinata*		1	2	1		4
Moraceae	*Artocarpus anisophyllus*			2			2
	A. elasticus				1x		1x
	A. lowii				2x		2x
	A. nitidus var. *griffithii*			2	1, 1x		3, 1x
	A. scortechinii			1	1x		1, 1x
	Ficus chartacea			1			1
Myristicaceae	*Gymnacranthera forbesii*				1		1
	Horsfieldia brachiata			2			2
	H. superba			1			1
	Myristica maingayi			1			1
Myrsinaceae	*Ardisia pachysandra*			1x			1x
Myrtaceae	*Eugenia griffithii*				1x		1x
	E. nigricans			1			1
	E. prianiana			1			1
	E. pseudocrenulata			1			1
	E. ridleyi			1, 1x			1, 1x
	E. tumida			1			1
	Eugenia sp.			1			1
Oxalidaceae	*Sarcotheca griffithii*			1			1
Polygalaceae	*Xanthophyllum eurhynchum*		1	1	1x		2, 1x
	X. stipitatum		1	1			2
Chrysobalanaceae	*Atuna excelsa*			1	1		2
Rubiaceae	*Nauclea officinalis*			1			1
	Porterandia anisophylla			4			4
Rutaceae	*Euodia glabra*			1x	1		1, 1x
	E. roxburghiana					1	1
Sapindaceae	*Nephelium costatum*			1			1
	N. hamulatum			1			1
	Pometia pinnata var. *alnifolia*			1	1		2
	Xerospermum noronhianum			2	3x		2, 3x
Sapotaceae	*Ganua* sp.			1	1, 1x		2, 2x

Continued

Table 3.3. *(Continued).*

Family	Species	1995		1996			
		Spring	Autumn	Spring	Autumn	1997	Total
	Payena lucida			3	1, 1x		4, 1x
Sterculiaceae	*Heritiera simplicifolia*					1	1
	Pterocymbium tubulatum			2			2
	Scaphium macropodum			1	1x	1	2, 2x
Styracaceae	*Styrax benzoin*					1, 1x	1, 1x
Symplocaceae	*Symplocos crassipes*			1			1
Thymelaeaceae	*Aquilaria malaccensis*			1			1
	Gonystylus maingayi			3			3
Tiliaceae	*Grewia blattaefolia*			1			1
	G. miqueliana				1x		1x
	Schoutenia accrescens			1			1
Ulmaceae	*Gironniera nervosa*			1			1
	G. subaequalis				1		1
Verbenaceae	*Teijsmanniodendron coriaceum*	1					1
Pandanaceae	*Galearia main gayi*				1, 1x		1, 1x
Fagaceae	*Castanopsis megacarpa*			1, 1x			1, 1x
	Lithocarpus curtisii			1			1
	L. rassa			2	2		4
Dipterocarpaceae	*Dipterocarpus cornutus*		1	2, 1x	2x		3, 3x
	D. kunstleri			2			2
	Neobalanocarpus heimii	1					1
	Parashorea densiflora				1x	1x	2x
	Shorea acuminata			3	1x		3, 1x
	S. bracteolata			1x			1x
	S. lepidota		3	5, 2x	9, 1x		17, 3x
	S. leprosula	2x	3	14		2	19, 2x
	S. macroptera			9, 1x	3, 4x		12, 5x
	S. maxwelliana			2	1		3
	S. multiflora	1		2			3
	S. ovalis	2			1, 1x		3, 3x
	S. parvifolia	2		6, 1x	2, 3x		10, 4x
	S. pauciflora			1		2, 2x	3, 3x
	Vatica bella					1x	1x

Note: x indicates that the collection of basidiomes was located between two or three trees, e.g. 2x indicates two collections of basidiomes each of which was located between two or three trees.

of putative ectomycorrhizal basidiomycetes, particularly with members of the *Dipterocarpaceae*, with at least 85 distinct collections (Table 3.3), is further reinforced here.

In Kalimantan, Indonesia, some host species of largely arbuscular mycorrhizal families have been reported to form ectomycorrhizal associations (Smits, 1992), but at present we know very little about the root structure and symbiotic associations of many of the tree species recorded here.

Conclusions

For the first time quantitative data have been obtained to ascertain the distribution, occurrence and frequency of putative ectomycorrhizal basidiomycetes in the South-east Asian rainforest context. Putative ectomycorrhizal fungi collected over a 2.5 year period in five 1 ha plots in a logged-over, regenerating lowland dipterocarp forest in Peninsular Malaysia have shown that the mycota in such communities is more diverse than previously accepted, even only a few years ago. Several hypogeous taxa have been recorded showing that this life-form is an important part of the tropical rainforest of South-east Asia. In any year, there are two distinct periods in the occurrence of basidiomes, but 'flush' years, where there is a large abundance of basidiomes, occur sporadically; both are probably related to climatic changes. Overall, the results did not indicate a similarity between the occurrence of putative ectomycorrhizal fungi in the plots, indicating a reticulate pattern of distribution of basidiomycetes even in a single community. There was a greater similarity between the species which fruited in spring and autumn than is generally found in boreal forests. The distribution of putative ectomycorrhizal basidiomycetes within the study area is undoubtedly linked to the previous history of land use and forest management methods.

Members of the *Russulaceae*, especially the genus *Russula*, made up the greatest number of colonies and generally numbers of basidiomes. Although members of other families appeared in significant numbers in some years, some of these did not occur every year. Suspected ectomycorrhizal members of the *Cortinariaceae* were confirmed to be rare, unlike those in boreal forests. Many new species in a range of families were recognized during the present study indicating our poor knowledge of tropical rainforest systems. The observations also appeared to identify new possible basidiomycete/tree associations in families of phanerogams not previously considered to be ectomycorrhizal. It is imperative that the roots of some of the tree families pinpointed in the present study are investigated to prove whether the associations based on basidiome occurrence are indeed correct. This will provide ultimately a better understanding of their significance in the overall mycorrhizal equation and their role in the forest ecosystem.

It is clear from the present study that a 2.5 year period of collecting in a tropical environment is insufficient to provide any true indication of the fungal fruiting patterns and diversity present in a complex lowland rainforest ecosystem.

Acknowledgements

The authors wish to thank their FRIM colleagues for logistical assistance and tree identication, Dr A. Taylor, Skoglig mycologi och pathologi, Uppsala, Sweden, for help in data gathering and the Natural Resources Institute, Chatham, for financial and administrative support.

References

Ahmad, N. (1989) Mycorrhizas in relation to Malaysian forest practice: a study of infection, inoculum and hosts response. PhD thesis, University of Aberdeen, UK.

Alexander, I.J. and Högberg, P. (1986) Ectomycorrhizas of tropical angiospermous trees. *New Phytologist* 102, 541–549.

Chipp, T.F. (1921) A list of the fungi of the Malay Peninsula. *Gardens' Bulletin, Straits Settlements* 2, 311–418.

Corner, E.J.H. (1935) The seasonal fruiting of agarics in Malaya. *Gardens' Bulletin, Straits Settlements* 9, 79–88.

Corner, E.J.H. (1940) Larger fungi in the tropics. *Transactions of the British Mycological Society* 24, 357–358.

Corner, E.J.H. (1994) Agarics in Malesia Part I Tricholomatoid. *Nova Hedwigia* 109, 1–138.

Grainger, J. (1957) Presidential address. From the field to laboratory and back. *The Naturalist (London)* 1957, 93–96.

Harley, J.L. and Smith, S.A. (1983) *Mycorrhizal Symbiosis*. Academic Press, London.

Hawksworth, D.L. (1993) The tropical fungal biota: census, pertinence, prophylaxis and prognosis. In: Isaac, S., Frankland, J.C., Watling, R. and Whalley, A.J.S. (eds) *Aspects of Tropical Mycology*. Cambridge University Press, Cambridge, pp. 265–293.

Henderson, D.M., Orton, P.D. and Watling, R. (1969) *British Fungus Flora: Agarics and Boleti: Introduction*. HMSO, Edinburgh.

Lee, S.S. (ed.) (1995) *A Guidebook to Pasoh*. FRIM Technical Information Handbook No. 3, FRIM, Kepong, Malaysia.

Lee, S.S. (1998) Root symbiosis and nutrition. In: Appanah, S. and Turnbull, J.M. (eds) *A Review of Dipterocarps: Taxonomy, Ecology and Silviculture*. CIFOR, Bogor, Indonesia, pp. 99–114.

Lee, S.S., Alexander, I.J. and Watling, R. (1997) Ectomycorrhizas and putative ectomycorrhizal fungi of *Shorea leprosula* Miq. (Dipterocarpaceae). *Mycorrhiza* 7, 63–81.

Lee, S.S., Watling, R. and Turnbull, E. (2002) Diversity of putative ectomycorrhizal fungi in Pasoh Forest Reserve. In: Okuda,T. (ed.) *Pasoh: Ecology and Natural History of a Southeast Asian Lowland Tropical Rainforest* (in press).

Louis, I. and Scott, E. (1987) *In vitro* synthesis of mycorrhiza in root organ culture of a tropical dipterocarp species. *Transactions of the British Mycological Society* 88, 565–568.

Manokaran, N. (1998) Effect, 34 years later, of selective logging in the lowland dipterocarp forest at Pasoh, Peninsular Malaysia, and implications on present-day logging in the hill forests. In: Lee, S.S., Dan, Y.M., Gauld, I.D. and Bishop, J. (eds) *Conservation, Management and Development of Forest Resources, Proceedings of the Malaysia–United Kingdom Programme Workshop, 21–24 October 1996, Kuala Lumpur*. FRIM, Kepong, Malaysia, pp. 41–60.

Noor, Nur Supardi Md., Dransfield, J. and Pickersgill, B. (1998) Preliminary observations on the species diversity of palms in Pasoh Forest Reserve, Negri Sembilan. In: Lee, S.S., Dan, Y.M., Gauld, I.D. and Bishop, J. (eds) *Conservation, Management and Development of Forest Resources. Proceedings of the Malaysia–United Kingdom Programme Workshop, 21–24 October 1996, Kuala Lumpur*. FRIM, Kepong, Malaysia, pp. 105–114.

Richardson, M.J. (1970) Studies on *Russula emetica* and other agarics in a Scots pine plantation. *Transactions of the British Mycological Society* 55, 217–229.
Singh, K.G. (1966) Ectotrophic mycorrhiza in equatorial rain forests. *Malay Forester* 29, 13–18.
Smits, W.T.M. (1992) Mycorrhizal studies in dipterocarp forests in Indonesia. In: Read, D.J., Lewis, D.H., Fitter, A.H. and Alexander, I.J. (eds) *Mycorrhizas in Ecosystems.* CAB International, Wallingford, UK, pp. 283–292.
Trappe, J.M. (1962) Fungus associates of ectotrophic mycorrhizae. *Botanical Review* 28, 538–606.
Turnbull, E. and Watling, R. (1999) Some records of *Termitomyces* from Old World Rainforests. *Kew Bulletin* 54, 731–738.
Watling, R. (1995) Assessment of fungal diversity: macromycetes the problems. *Canadian Journal of Botany* 73 (Suppl.) 1, S15–S24.
Watling, R., Lee, S.S. and Turnbull, E. (1998) Putative ectomycorrhizal fungi of Pasoh Forest Reserve, Negeri Sembilan. In: Lee, S.S., Dan, Y.M., Gauld, I.D. and Bishop, J. (eds) *Conservation, Management and Development of Forest Resources. Proceedings of the Malaysia–United Kingdom Programme Workshop, 21–24 October 1996, Kuala Lumpur.* FRIM, Kepong, Malaysia, pp. 96–104.

Basidiomycetes of the Greater Antilles Project 4

D.J. Lodge,[1] T.J. Baroni[2] and S.A. Cantrell[1]

[1]Center for Forest Mycology Research, Forest Products Laboratory, USDA-Forest Service, PO Box 1377, Luquillo, PR 00773-1377, USA; [2]Department of Biological Sciences, SUNY Cortland, PO Box 2000, Cortland, NY 13045, USA

Introduction

The inventory of basidiomycetes of the Greater Antilles, with special emphasis on the Luquillo Long-Term Ecological Research Site, was a 4 year project initiated in 1996 with funding from the USA National Science Foundation's (NSF) Biotic Surveys and Inventories Program.* The objective was to survey and inventory all basidiomycetes except rust fungi on the Caribbean islands of Puerto Rico, the Virgin Islands, Hispaniola and Jamaica. Although Cuba was not visited, historical records from there have been included in the analyses.

Basidiomycetes, except for those with durable basidiomes such as *Lentinus* spp. and polypore fungi, were poorly known in the Greater Antilles before the project began. The Caribbean was colonized five centuries ago and many botanists of early date collected hard polypores since the basidiomes retained their characteristics sufficiently for later identification. Although Stevenson (1970) had summarized all the previous records of fungi from Puerto Rico and the nearby Virgin Islands, he listed only 55 species of ephemeral basidiomycetes (Lodge, 1996a). Lodge (1996b) added 170 ephemeral basidiomycetes to the list for Puerto Rico, of which 10–30% from various families were previously undescribed species. Ciferri (1929) listed only 14 species of agarics and nine species of gasteromycetes among the 63 macrobasidiomycete species known from the

*The Forest Products Laboratory in Madison is maintained in cooperation with the University of Wisconsin, while the laboratory in Puerto Rico is maintained in cooperation with the USDA-Forest Service International Institute of Tropical Forestry. This article was written and prepared by a US Government employee on official time, and the information is therefore in the public domain and not subject to copyright.

Dominican Republic on the island of Hispaniola, whereas 35 were species of polypores and other aphyllophoroid fungi. Benjamin and Slot (1969) summarized the fungi known from Haiti on Hispaniola based on specimens deposited at Beltsville, Maryland, including two species not recorded by Ciferri (1929). A parataxonomist recently added 19 new records of agaric species from the Dominican Republic (Rodríguez Gallart, 1989, 1990, 1997) but corresponding voucher collections were not located in the National Herbarium.

The basidiomycete fungi in Jamaica and Cuba had received somewhat more attention than those in Puerto Rico and Hispaniola. Swartz (1788) collected and described several species of polypores from Jamaica, and Hennings (1898) described three more species collected by Lindau. Murrill (1910, 1911a, b, c, d, 1915, 1920a,b, 1921a,b) collected in Jamaica in the early 1900s and described many new species. In addition, Dennis (1950, 1953) had made one bountiful collecting trip to Jamaica. Ryvarden (1985) revised the polypores described by Murrill, and the *Hygrophoraceae* were included in Hesler and Smith's (1963) monograph. Despite these efforts, the basidiomycetes of Jamaica remained poorly known, many represented by a single collection. Historically, the basidiomycetes of Cuba had received almost as much attention as those in Jamaica. Collections by Charles Wright were described by Berkeley and Curtis (1868) and revised by Murrill (1911a, b, c, d); Dennis (1950, 1951a, b, c); Pegler (1983), Ryvarden (1984) and Hjortstam (1990). Earle (1906) collected widely in Cuba in the early 1900s; he described some species in 1906 and 1909, but sent many others on to W.A. Murrill, who revised and described them with other West Indian, Honduran and Mexican collections (1911a, b, c, d).

Investigators in the course of this survey and inventory of basidiomycetes in the Greater Antilles have discovered at least 75 new species and varieties so far, as well as several new genera and one possible new family or order. In this chapter, the percentages and numbers of new taxa are compared with original estimates. Problems encountered in classifying some of the collections, and some surprising results from DNA analyses that were used to resolve their placement are also discussed. Some of the biogeographic patterns that have emerged are also shown.

Predicted versus Observed New Species

Methods for generation of predicted values

Predictions of the number of new species that have yet to be discovered are generally based on the previous rate of discovery. The percentage of expected species in each family was assumed to be the same as in the recent data from Puerto Rico (1983–1995; Lodge, 1996b). The expected total number of species for each family was calculated using the proportional overlap in species composition between what was known in the Greater Antilles and the better known mycotas from south-eastern USA, the Lesser Antilles (Pegler, 1983), Venezuela and Trinidad (Dennis, 1953,1970), multiplied by the number of species reported from those areas.

Duplications were eliminated in cases where the same species occurred in both comparative mycotas before the expected number of species were tallied. In the *Hygrophoraceae* for example, 17 species were known from Puerto Rico (those in Lodge, 1996b except *Hygrocybe unicolor* Pegler) of which one was known from south-eastern USA (Hesler and Smith, 1963). The monograph of Hesler and Smith (1963) listed 49 species of *Hygrocybe*, *Camarophyllopsis* and *Cuphophyllus* found in south-eastern USA, excluding species in common with the Lesser Antilles, Trinidad and Venezuela. It was expected that 5.9% (1/17) of these 49 would eventually be found in the Greater Antilles, namely two in addition to the one species already known. For the Lesser Antilles, Trinidad and Venezuela 38 species are known (Dennis, 1970; Pegler, 1983), of which ten were recorded from Puerto Rico (Lodge, 1996b). It was expected that 58.8% (10/17) of the 38 southern Caribbean species would eventually be found in the Greater Antilles, namely 12 in addition to the ten already known. A total of 23 species of *Hygrophoraceae* were known from the Greater Antilles: one (*Hygrocybe cantharellus* (Schw.) Murrill) from Jamaica and south-eastern USA (Dennis, 1953; Hesler and Smith, 1963), five known only from Jamaica or Cuba (Murrill, 1911c; Hesler and Smith, 1963) and 17 from Puerto Rico (Lodge, 1996b). The expected total number of described species of 37 was obtained as the tally of 23 species already recorded from the Greater Antilles, an additional two from North America and an additional 12 from the southern Caribbean. The percentage of expected new species was assumed to be the same as previously (23.5%). Thus, the total number of expected species ($X = 48$) and the number of undescribed species ($Y = 11$) were estimated by solving the following simultaneous equations:

$$37 \text{ known species} + \text{undescribed species } (Y) = \text{total number of species } (X) \quad (1)$$

$$Y/X = 0.235 \quad (2)$$

The expected numbers of species (total and undescribed) in the *Entolomataceae* for the Greater Antilles were generated from species already recorded (Berkeley and Curtis, 1868; Murrill, 1911d; Hesler, 1967; Lodge, 1996b) and the comparative mycotas in south-eastern North America (Hesler, 1967) and the Lesser Antilles, Trinidad and Venezuela (Dennis, 1970; Pegler, 1983). The expected number of polypores (*sensu lato*) was generated in the same way, using polypore mycotas in Venezuela (Dennis, 1970) and south-eastern North America (Gilbertson and Ryvarden, 1987), but the percentage of undescribed species was based on an ‘expert guess’.

Results

Total numbers of species and new species

Progress in classifying the nearly 5000 collections generated by the project varied among families. Therefore, total numbers of species and new species could be compared with original predictions only for polypores, *Hygrophoraceae* and *Entolomataceae* (Table 4.1). Several publications documenting additions to the mycota of the Greater Antilles followed as a result of this project, including polypores (Decock and Ryvarden, 2000; Ryvarden, 2000a, b, c, 2002), the *Corticiaceae* (Lodge, 1996b; Nakasone *et al.*, 1998), the *Hygrophoraceae* (Cantrell and Lodge, 2000, 2001), the *Entolomataceae* (Baroni and Lodge, 1998), and the *Amanitaceae*, *Boletaceae* and *Russulaceae* (Miller *et al.*, 2000).

The number of species of polypores found in the Greater Antilles was 22% more than the original estimate (Table 4.1). While at least ten unexpected representatives of African or palaeotropical species were found, the latter represented only a quarter to a third of the 38 polypore species that were in excess of the original estimate (Table 4.1). Species that are common in Africa or Asia and are rarely represented in the Greater Antilles and the Neotropics included *Abundisporus fuscopurpureus* (Jungh.) Ryv., *Cerrena meyenii* (Kl.) Hansen, *Flavodon flavus* (Kl.) Ryv., *Fomitopsis dochmius* (Berk.) Ryv., *Lenzites acuta* Berk., *Navisporus floccosus* (Bres.) Ryv., *Trametes cingulata* Berk., *Trichaptum byssogenus* (Jungh.) Ryv. and *Trichaptum durum* (Jungh.) Corner. All but one of the previously undescribed species of polypores from the Greater Antilles had ephemeral basidiomes. However, fewer new polypore species were found than expected (Table 4.1), so this cannot explain why more species of polypores were found than predicted. One contributing factor in the original underestimate could have been that the reference mycotas were incomplete. The plot-based methods which are more commonly used to project the total number of species in an area also produce underestimates (Schmit *et al.*, 1999).

The number of species in the *Hygrophoraceae* found in the Greater Antilles was also 40% more than estimated using the same methods as for polypores (Table 4.1), and the number of undescribed species was slightly higher than expected (Table 4.1), but this can explain only six of the 17 species in excess of the authors' original estimate for the *Hygrophoraceae*. The percentages of undescribed species recorded before (23.5%; Lodge and Pegler, 1990; Lodge, 1996b) and after the Basidiomycetes of the Greater Antilles Project (26%; Cantrell and Lodge, 2000a, b) were not very different. Thus, it appears that while the methods used yielded predictions for the total number of species that were in the right range, they were lower than the observed values.

Table 4.1. Predicted versus observed total numbers of species and numbers of previously undescribed species in the Greater Antilles.

Group	Total number of species predicted	Total number of species found	Expected number of new species	Observed number of new species
Polypores	170	208	10	6
Hygrophoraceae	48	63	11	17
Entolomataceae	88	89+	30	28+

Percentage of undescribed species

The percentages of new species before and after the inventory began could be compared for polypores, agarics and boletes in Puerto Rico based on partial results (Table 4.2). Only taxa that were classified to species or listed as undescribed species were included in this analysis, so the total number of agaric species for the island are likely to be at least two to five times higher. The predicted percentages of undescribed species were remarkably similar to the observed percentages in Table 4.2.

The percentage of undescribed species revealed in a survey could be expected to decline eventually as an inventory for the area nears completion. It was thought, therefore, that the percentage of undescribed species would decrease from the percentage observed in the 12 years prior to the project. This did not occur, suggesting that the inventory was not complete. The tendency for mycologists to work first on the obviously new and interesting species may bias the percentages of undescribed species and keep them unnaturally high. A significant decline in the percentage of new species, however, did not occur even among the groups of agarics that had been analysed completely (i.e. *Hygrophoraceae* 27%; and in the *Entolomataceae*, *Alboleptonia* 45% and *Pouzarella* 83%).

Problems and Surprising Results in Systematics

Several different types of problem were encountered in classifying the basidiomycetes collected in the survey of the Greater Antilles. Some of the problematic species were well known but were obviously placed in an incorrect genus. In addition, some of the new species were readily placed in existing genera but had no obvious close relatives. Other fungi were taxonomically anomalous, and belonged to none of the previously described genera.

Species previously placed in the wrong genus

For some of the previously named species found in the Greater Antilles that had been placed in the wrong genus, it was not immediately obvious where they

Table 4.2. Predicted versus observed percentages of previously undescribed species in Puerto Rico before and after the Basidiomycetes of the Greater Antilles Project. Predicted percentages of new species in the *Agaricales* and *Boletales* taken from the rate of discovery of new species in the preceding 10 years.

Group	Total number of species determined	% undescribed species	Predicted % undescribed species
Agaricales and *Boletales*	325	22	22
Polypores, *sensu lato*	118	4	6

belonged on the basis of morphological characteristics alone, but ribosomal DNA sequencing by R. Vilgalys and J.-M. Moncalvo (see Acknowledgements) was instrumental in classifying them correctly. One such example was *Collybia aurea* (Beeli) Pegler, a gregarious, lignicolous species that was originally described in the genus *Marasmius*. The pileipellis of this brilliant mustard yellow species is a cutis of repent hyphae with scattered, simple, clavate, upturned terminal elements (Pegler, 1983). Neither the coloration (Legon, 1999, Fig. 6) nor the pileipellis structure is typical of either *Marasmius* or *Collybia*, or of any of the segregate genera. Vilgalys and Moncalvo found that the ribosomal DNA sequences of *C. aurea* lined up with species in the genus *Tricholomopsis* (R. Vilgalys, Durham, North Carolina, USA, 1999, personal communication). In retrospect, this placement was logical, but it was not a solution that was even considered before the DNA sequences resolved the problem. *Marasmius rhyssophyllus* Mont. (syn. *Dictyoploca rhyssophylla* (Mont.) Baker & Dale) was another misclassified species resembling *C. aurea* in colour, habit, and pileipellis structure but differing in distant and highly intervenose lamellae (Legon, 1999, Fig. 7). It also belonged in *Tricholomopsis* according to the ribosomal DNA evidence (J.-M. Moncalvo, Durham, North Carolina, USA, 2000, personal communication).

Another species in the West Indies that had clearly been placed in the wrong genus and was previously referred to as *Tricholoma pachymeres* (Berk. & Br.) Sacc. was provisionally identified as saprobic *Tricholoma titans* H.E. Bigelow & Kimbr. The molecular data, microscopic characters, especially the lack of siderophilous granulation in the basidia and the abundance of clamp-connections, and ecological information led to the erection of *Macrocybe* Pegler & Lodge for this and related tropical species (Pegler *et al.*, 1998). More recent work indicated that *Macrocybe* did not cluster in either the *Tricholoma* or *Calocybe* clades (Moncalvo *et al.*, 2000).

Expanding the limits of described sections, subgenera and genera

The most interesting ectomycorrhizal species found was an *Amanita* in the subgenus *Amanita*. It is characterized by having two layers of universal veil that differ in colour and texture. The outer layer is thin, white and membranous with large pyramidal warts situated in the sculptured depressions of the pileus. The inner universal veil is powdery and rusty brown, covering the entire basidiome and separating from the lower side of the partial veil as a false annulus (Cantrell *et al.*, 2001, Fig. 6). The powdery veil also sloughs off the surface of the pileus, which becomes shiny and has an orange tint. There are some species with composite universal veils in the subgenus *Lepidella*, but the two veil layers are usually the same colour. O.K. Miller Jr (Blacksburg, Virginia, USA, 2000, personal communication) has not found any species in subgenus *Amanita* that is closely related to this new *Amanita* from the Dominican Republic (Miller and Lodge, 2001). *Amanita* species from the Dominican Republic have been described by Miller and Lodge (2001).

The authors are describing a new species of *Callistodermatium* Singer. This previously monotypic genus was described by Singer from South America, and differs from a related genus, *Cyptotrama*, in having pileus pigments that change colour with alkaline solutions, somewhat larger spores, and a terrestrial rather than lignicolous habit. While the type species *Callistodermatium* is dull brown, the new species found in Puerto Rico and the Dominican Republic has a pileus that was bright orange-yellow, turning fuchsia pink rather than violet in KOH (see Cantrell *et al.*, 2001, Fig. 2).

The limits of *Hygrocybe* section *Firmae* have been greatly altered by the new species discovered during the Basidiomycetes of the Greater Antilles Project (Cantrell and Lodge, 2001). Two of the new species, *Hygrocybe cinereofirma* S.A. Cantrell, Lodge, & Baroni, and *Hygrocybe brunneosquamosa* Lodge & S.A. Cantrell are dull grey and grey-brown in colour, which is very unusual for section *Firmae*. All the previously known members of this section have at least some red, yellow, green or purple coloration. *H. brunneosquamosa* is also the only member of the section with squarrose scales on the pileus (Cantrell *et al.*, 2001, Fig. 3). In addition, all the previously described species in section *Firmae* have a broadly convex or centrally depressed pileus, whereas three of the new species (*H. cinereofirma*, *Hygrocybe flavocampanulata* S.A. Cantrell & Lodge, and *Hygrocybe laboyi* S.A. Cantrell & Lodge) have a pileus disk that is umbonate or cuspidate. Previously, all the species in section *Firmae* except *Hygrocybe hypohaemacta* (Corner) Pegler had broadly attached lamellae; in this project, two new species with strongly adnexed lamellae (*H. flavocampanulata*, and *H. cinereofirma*; Cantrell and Lodge, 2001) were described.

Interesting new sections, genera and families

One of the most exciting finds on the island of Jamaica was a gregarious, greenish blue mycenoid species identified as *Clitocybula azurae* Singer (Cantrell *et al.*, 2001, Fig. 7). The dextrinoid stipe context, slightly dextrinoid lamellar trama hyphae, umbonate rather than umbilicate shape of the pileus, and the unusual greenish blue colour of the basidiomes in *C. azurae* are characteristics that are not shared with typical species of *Clitocybula*, which suggested that *C. azurae* was misplaced. *C. azurae* resembles a *Mycena* species in having amyloid spores, dextrinoid stipe tissue, and a mycenoid structure in the lower pileus and upper lamellar contexts, but the unornamented pileipellis hyphae and cheilocystidia, greenish blue rather than purplish pigments, absence of a mycenoid structure in the hypoderm, and absence of a separation zone between the stipe and pileus context suggested that it should not be placed in the genus *Mycena*. It is uncertain where this unusual species will eventually be classified. Many of the undescribed true *Mycena* species from the Caribbean also do not fit into any of the currently described sections. Another agaric with unknown affinities is a species called the 'nail-head fungus' because of its shape, dark colours and tough texture (Cantrell *et al.*, 2001, Fig. 9). Although this species somewhat resembles *Tephrocybe*, it lacks siderophilous granulation in the basidia, and therefore appears to belong in the tribe *Tricholomatae* rather than *Lyophyllae*, and may be an undescribed genus.

A new ochraceous coloured species of *Gloeocantharellus* was discovered on the island of Tortola in the British Virgin Islands. The ribosomal DNA sequences showed that it was basal to the *Gomphaceae*, consistent with *Gloeocantharellus* (J. Spatafora, Liverpool, UK, 2000, personal communication). Another unusual discovery was an undescribed *Dichopleuropus*-like fungus (D. Reid, West Sussex, UK, 1999, personal communication). Its basidiomes were flabelliform, somewhat fleshy, caespitose or forked above the base, and growing from roots; microscopically, this had gloeocystidia, and spores that were hyaline, faintly amyloid, broadly ellipsoid or subglobose, and smooth (Lodge *et al.*, 2001, Fig. 12). This previously undescribed species differs from *Dichopleuropus* in lacking dextrinoid dichophyses in the tramal tissues, which is a major distinguishing feature of *Dichopleuropus*. Sequencing of the nuclear DNA 5.8S, ITS2, and the 5′-end of the 28S (LSU) regions by K.-H. Larsson and E. Larsson (Goteborg, Sweden, 2000, personal communication) indicated that this fungus (PR-5100) had affinities with taxa in the *Lachnocladiaceae*, which is consistent with the hypothesized placement of *Dichopleuropus* (Boidin *et al.*, 1998).

DNA sequences did not resolve all taxonomic problems. The placement of a group of fungi that macroscopically resembled *Stereum*, *Thelephora* and *Dichopleuropus*, and typified by collections from both Puerto Rico and Venezuela, remains unresolved. Their nuclear DNA sequences did not match any of those of the 170 taxa covering the major homobasidiomycete groups studied by Larsson and Larsson (Goteborg, Sweden, 2000, personal communication).

Some of the fungi collected by the group included species other than basidiomycetes. The most notable of these was a new resupinate genus and species of ascomycete, *Rogersonia striolata* Samuels & Lodge, that was mistaken for a corticiaceous fungus in the field. The new genus was closely related to *Hypomyces* in its stromatal morphology, but there was no indication of a fungal host, the apical discharge mechanism was absent from the asci, the spores were single-celled and broadly ellipsoid rather than two-celled, fusiform and apiculate; and the transverse striations on the ascospores of *Rogersonia* are otherwise unknown in the *Hypocreales*. Other additions to the *Hypocreaceae* included several new parasitic species of *Hypomyces* that were collected on polypores (Poldmaa *et al.*, 1997). An undescribed species of *Camarops* (*Xylariaceae*) which changed the normally grey pore surface yellow was found on vigorously growing *Tinctoporellus epimiltinus* (Berk. & Br.) Ryvarden.

Biogeographic Affinities

Some intriguing biogeographic patterns emerged from the agaric families that were examined most thoroughly (Lodge *et al.*, 2001). These patterns gave some indication of centres of origin or speciation for certain genera, subgenera and sections. The two families that received the most attention were the *Hygrophoraceae* and *Entolomataceae*.

Hygrophoraceae

More than one-third of the 66 species and varieties identified in the *Hygrophoraceae* have been found only in the Greater Antilles (Lodge *et al.*, 2001; Table 4.3). A quarter of the species and varieties of *Hygrophoraceae* in the Greater Antilles have also been reported by Pegler (1983) from the Lesser Antilles. Only a few species are restricted to the Caribbean islands (both Greater and Lesser Antilles), but 17% are found scattered throughout the Caribbean Basin (Mexico, Central America, northern South America, and the Greater and Lesser Antilles; Table 4.3). A smaller percentage of species extend far into South America through the Lesser Antilles, or are absent from the Lesser Antilles (Table 4.3). The Neotropical elements in this family are represented primarily by *Hygrocybe* species in sections *Firmae* and *Coccineae*, and two species of *Hygroaster*. The Neotropics may be a centre of recent speciation for section *Firmae* since the greatest number of species are found there (Cantrell and Lodge, 2001; Lodge *et al.*, 2001).

Section *Firmae* is characterized by having dimorphic spores and basidia. Only one species in Section *Firmae* is pantropical, *H. hypohaemacta* (Corner) Pegler; it is also the only species that is found in moist as well as wet forests and as far north as Texas. At least three species in section *Firmae* are known from Mexico, including *H. hypophaemacta* (J. Garcia, Ciudád Victória, 2000, personal communication). It is significant that there has been only one possible report of a species in section *Firmae* from Costa Rica (C. Ovreboe, Edmond, Oklahoma, 2000, personal communication) despite extensive collecting there. Costa Rica and neighbouring parts of Central America are geologically young (approximately 3 million years old), whereas Mexico, parts of the Antilles and South America are much older. Their rarity in Central America suggests that species in section *Firmae* generally have poor dispersal and colonizing abilities, which may contribute to genetic isolation among populations on different islands or in different regions, and subsequent speciation.

Other taxa in the *Hygrophoraceae* appear to have a temperate origin, such as species in *Hygrocybe* subgenus *Hygrocybe* (Lodge *et al.*, 2001). Cantrell and Lodge (2000) described two new varieties of temperate species in this subgenus, *Hygrocybe calyptriformis* var. *domingensis* S.A. Cantrell & Lodge, and *Hygrocybe konradii*, var. *antillana* S.A. Cantrell & Lodge. These species may have reached the Caribbean recently. Species in subgenus *Hygrocybe* may disperse and colonize better than species in subgenus *Pseudohygrocybe*, section *Firmae*. While the Caribbean populations of *H. calyptriformis* and *H. konradii* differ by one or two morphological characters from their temperate relatives, greater dispersal and

Table 4.3. Geographic ranges are given for 63 species of *Entolomataceae* and 66 species and varieties of *Hygrophoraceae* that have been identified from the Greater Antilles. The number and percentage of species known from a given geographic range are presented (GA, Greater Antilles; LA, Lesser Antilles; SA, South America, and NA, North America). The Caribbean Basin includes Central America, northernmost South America (Trinidad and Venezuela) and the Antilles.

Range	Species of *Entolomataceae*	% of all species *Entolomataceae*	Species and varieties of *Hygrophoraceae*	% of all species *Hygrophoraceae*
GA	26	41.3	25	37.9
GA and LA	8	12.7	2	3.0
Caribbean Basin	10	15.9	11	16.7
GA, LA, and SA	0	0	7	10.6
GA and SA	3	4.8	4	6.13
Africa	34	4.8	0	0
Pantropical	2	3.2	1	1.56
New World	0	0	0	0
NA	9	14.3	2	3.0
NA and Europe	2	3.2	5	7.6
N. Temperate	0	0.0	4	6.1
Worldwide	0	0	5	7.6

colonization abilities in this subgenus may preclude the degree of genetic isolation needed to form a separate species. Other species with otherwise north temperate distributions also appear in the Greater Antilles, for example, *Hygrocybe caespitosa* Murrill and *Hygrocybe ovina* (Fr.) Kühner from North America, and *Hygrocybe* cf. *mucronella* (Fr.) P. Karst. from Europe, the latter differing from European collections in having broader spores.

Entolomataceae

Almost half the 63 species of *Entolomataceae* examined were apparently restricted to the Greater Antilles (Table 4.3). While species in various genera in this group were represented, there was a preponderance of *Alboleptonia* (Baroni and Lodge, 1998), *Claudopus* and *Pouzarella* species with highly restricted distributions (Lodge *et al.*, 2001). In contrast, three of the five *Rhodocybe* species were scattered throughout the Caribbean Basin. Less than a third (29%) of the species of *Entolomataceae* in the Greater Antilles were recorded by Pegler (1983) from the Lesser Antilles. While only a few species of *Entoloma* and *Leptonia* are found in both the Greater Antilles and South America below the Caribbean Basin (e.g. *Entoloma dragonosporum* (Singer) Horak and *Entoloma lowyi* (Singer)

Horak; Table 4.3), the closest apparent relatives of many species in the Greater Antilles are found in South America (Baroni *et al.*, 1997), and only two species, *Alboleptonia stylophora* (Berk. & Br.) Pegler and *Entoloma virescens* (Berk. & Curt.) Horak, have pantropical distributions. Neotropical collections of *A. stylophora*, however, are strikingly paler than the yellow forms found in the palaeotropics (E. Horak, Jamaica,1999, personal communication). Only a few species, such as *Inocephalus murraii*, *Leptonia incana* (Fr.) Gill and *Pouzarella foetida* Mazzer extend from North America or the North Temperate Zone into the Greater Antilles, but some eastern North American species have a sibling species in the Greater Antilles (Baroni *et al.*, 1997). The most surprising disjunct distributions found in the *Entolomataceae* in the Greater Antilles were those of *Inocephalus lactifluus* and *Inopilus inocephalus* from Madagascar (Baroni *et al.*, 1997). The pattern of disjunct populations of species and sibling species pairs in East Africa or Madagascar and the Caribbean or eastern Neotropics has similarly been found among ferns (Moran and Smith, 1999).

Ectomycorrhizal fungi

The island of Hispaniola has the easternmost extent of native pine in the Caribbean, *Pinus occidentalis* Swartz, which is an endemic species. Many of the ectomycorrhizal fungi found in the Dominican Republic on Hispaniola, for example *Strobilomyces confusus* Singer, are also found in south-eastern USA. Therefore, the presence of *Suillus albivelatus* Smith, Thiers & O.K. Miller among the collections was surprising, since it was known previously only from Idaho and the Pacific Northwest in the USA (Smith *et al.*, 1965; Bessette *et al.*, 1999). Collections from the Dominican Republic have been identified as *Lactarius rubrilacteus* A.H. Smith & Hesler, known previously from Washington to California, the Rocky Mountains from Idaho to New Mexico, and Mexico. Recent identifications of *Lactarius* collections from the Dominican Republic (A. Methven, Charleston, Illinois, 2000, personal communication) confirmed the occurrence of western North American disjuncts in the Greater Antilles. In addition to *L. rubrilacteus*, the western North American disjuncts included *Lactarius deliciosus* (Fr.) S.F. Gray var. *areolatus* A.H. Smith (Alaska to California, the Rocky Mountains from Idaho to New Mexico, and Mexico), and *Lactarius scrobiculatus* (Fr.) Fr. var. *canadensis* (A.H. Smith) A.H. Smith (Alaska east across Canada, south along the Pacific Coast to California, the Rocky Mountains from Idaho to Colorado, and Mexico) (Lodge *et al.*, 2001).

The possibility of an accidental introduction of western North American ectomycorrhizal fungi to Hispaniola cannot be ruled out without molecular data, but north-western North America is an unlikely source for an introduction, and the foresters in the Dominican Republic reportedly never brought rooted pines into the country. Alternatively, some of these western North American disjuncts might well be the remnants of ancient distributions that extended south through the Rocky Mountains into the mountains of Mexico, Belize, northern Honduras and Guatemala, and east to the Greater Antillean islands. The island of Hispaniola was closer to northern Central America 35–65 million years ago than it is now (Hedges, 1992), which could have facilitated colonization by ectomycorrhizal fungi from the mainland; but more recent dispersal (not more than 23–25 million years ago) to the islands is more likely to have occurred based on the palynological history of pine in the Caribbean (Lodge *et al.*, 2001).

Summary and Conclusions

Although classification of all the collections from the Basidiomycetes of the Greater Antilles Project was not complete at the time of writing, it was clear that original estimates of the number of species and previously undescribed taxa in the Greater Antilles were conservative. Molecular data did not resolve all the taxonomic problems, but they were of great assistance in placing most of the new, as well as previously described, species into the correct genus and family. Molecular data were also crucial in establishing new genera.

A surprisingly high proportion (between one-third and one-half) of the species of *Hygrophoraceae* and *Entolomataceae* were apparently restricted to the Greater Antilles, with relatively little overlap between the Greater and Lesser Antilles, and pantropical species were rare. The biogeographical origins of basidiomycetes in the Greater Antilles are varied, but include South American and eastern North American elements, as well as African and western North American disjuncts. Some of the disjunct populations may be remnants of ancient distributions, whereas some of the species complexes in the *Hygrophoraceae* appear to have diversified recently. It is not known how fast basidiomycete fungi evolve, but many of the reptiles and amphibians of the Antilles that were once thought to have evolved tens of millions of years ago have arisen instead since the ice ages tens of thousands of years ago (James D. Lazelle, Guana Isl. BVI,1999; S. Blair Hedges, Orcas Island, Washington, USA, 2000, personal communications).

Acknowledgements

This work was made possible by grant DEB-95–25902 to the State University of New York College at Cortland from the National Science Foundation, Biotic Surveys and Inventories Program, and a joint venture agreement with the USDA-Forest Service, Forest Products Laboratory. In addition to the authors, the primary research group included Drs J. Carranza, K.K. Nakasone, K.-H. Larsson, R.E. Halling, E. Horak, Orson K. Miller Jr, P. Roberts, L. Ryvarden, R. Vilgalys and Mr A. Nieves-Rivera. Many other institutions and non-governmental organizations provided matching funds and support which greatly facilitated the research, including the USDA-Forest Service's International Institute of Tropical Forestry (IITF), particularly the International Forestry Program, and NSF grant BSR-8811902 to the Terrestrial Ecology Division of the University of Puerto Rico and IITF in support of the Luquillo Experimental Forest Long Term Ecological Research Site in Puerto Rico. Molecular sequencing by R. Vilgalys and J.-M. Moncalvo, with funding from the National Science Foundation grants DEB-9708035 and DEB-9408347, 'The Agaricales: Molecular Systematics and Evolution of Mushrooms', was instrumental in classifying some of the unknown species. We thank our cooperators in the Dominican Republic, especially Sr Andrés Ferrer of the Fundación Moscoso Puello, Sra C. Cassanova and the staff of Fundación Plan Sierra, Fundación Progressio, Ing. Milciedes Mejía and Sra Daisy Castillo of the Jardín Botánico, Dr Raphael M. Moscoso, and the National Park Service. We also thank Ms Tracey Commock and the Institute of Jamaica, Natural History Division for assistance in Jamaica; Dr Carolyn Rogers of the USGS, the Maho Bay Ecotourist Resort, and the Virgin Islands National Park Service on St John, USVI; and the Conservation Agency and the Falconwood Foundation for support while working on Guana and Tortola Islands. We are eternally grateful to Ms Pat Catterfeld of the Research Foundation of SUNY-Cortland and L.A. Barley for taking care of a myriad of details in support of our research. Many people outside our research team assisted us with identifications (Drs D. Boertmann, R. Courtecuisse, G. and L. Guzmán, T. Laessøe, A. Methven, G. Mueller, D. Pegler, D. Reid and G. Samuels), and DNA sequencing (E. Larsson, J.-M. Moncalvo and J. Spatafora).

References

Baroni, T.J. and Lodge, D.J. (1998) Basidiomycetes of the Greater Antilles I: new species and new records of *Alboleptonia* from Puerto Rico and St John, USVI. *Mycologia* 90, 680–696.

Baroni, T.J., Lodge, D.J. and Cantrell, S.A. (1997) Tropical connections: sister species and species in common between the Caribbean and the eastern United States. *McIlvainea* 13, 4–19.

Benjamin, C.R. and Slot, A. (1969) Fungi of Haiti. *Sydowia* Ser.II., 23, 125–163.

Berkeley, M.J. and Curtis, M.A. (1868) Fungi Cubenses (Hymenomycetes). *Journal of the Linnean Society, Botany* 10, 280–392.

Bessette, A.E., Roody, W.C. and Bessette, A.R. (1999) *North American Boletes*. Syracuse University Press, Syracuse, New York.

Boidin, J., Mugnier, J. and Canales, R. (1998) Taxonomie moleculaire des *Aphyllophorales*. *Mycotaxon* 66, 445–491.

Cantrell, S.A. and Lodge, D.J. (2000) Hygrophoraceae (Agaricales) of the Greater Antilles. *Hygrocybe*, subgenus *Hygrocybe*. *Mycological Research* 104, 873–878.

Cantrell, S.A. and Lodge, D.J. (2001) *Hygrophoraceae* of the Greater Antilles, Section *Firmae*. *Mycological Research* 105, 215–224.

Cantrell, S.A., Lodge, D.J. and Baroni, T.J. (2001) Basidiomycetes of the Greater Antilles Project. *The Mycologist* (The Mycologist 15: 102–112).

Ciferri, R. (1929) *Micoflora Domingensis.* Estación Agronómica de Moca, Republica Dominicana, Seri B – Botanica No. 14, J.R. Vda. García, Sucesores, Santo Domingo, Dominican Republic.

Decock, C. and Ryvarden, L. (2000) Studies in Neotropical polypores. 6. New resupinate *Perenniporia* species with small pores and small basidiospores. *Mycologia* 92, 354–360.

Dennis, R.W.G. (1950) An earlier name for *Omphalina flavida* Maubl. & Rangel. *Kew Bulletin* 5, 434.

Dennis, R.W.G. (1951a) Some *Agaricaceae* of Trinidad and Venezuela. Leucosporae: Part 1. *Transactions of the British Mycological Society* 34, 411–480.

Dennis, R.W.G. (1951b) *Lentinus* in Trinidad B.W.I. *Kew Bulletin* 5, 321–333.

Dennis, R.W.G. (1951c) *Agaricaceae* referred by Berkeley and Montagne to *Marasmius, Collybia* and *Heliomyces. Kew Bulletin* 6, 387–410.

Dennis, R.W.G. (1953) Some West Indian collections referred to *Hygrophorus* Fr. *Kew Bulletin* 8, 253–268.

Dennis, R.W.G. (1970) *Fungus Flora of Venezuela and Adjacent Countries.* Kew Bulletin Additional Series 3, HMSO, London.

Earle, F.S. (1906) Algunos hongos cubanos. *Información Anual Estación Central Agronomica Cuba* I, 225–242.

Earle, F.S. (1909) Genera of North American gill fungi. *Bulletin of the New York Botanical Gardens* 5, 375–462.

Gilbertson, R.L. and Ryvarden, L. (1987) *North American Polypores*, Vols 1 and 2. Fungiflora, Oslo, Norway.

Hedges, S.B. (1992) Caribbean biogeography: molecular evidence for dispersal in West Indian terrestrial vertebrates. *Proceedings of the National Academy of Sciences USA* 89, 1909–1913.

Hennings, P. (1898) Fungi jamaicensis. *Hedwigia* 37, 277–282.

Hesler, L.R. (1967) Entoloma in southeastern North America. In: *Beihefte zur Nova Hedwigia*, Heft 23. J. Cramer, Stuttgart, Germany.

Hesler, L.R. and Smith, A.H. (1963) *North American Species of Hygrophorus.* The University of Tennessee Press, Knoxville, Tennessee.

Hjortstam, K. (1990) Corticoid fungi described by M.J. Berkeley: II. Species from Cuba (West Indies). *Mycotaxon* 39, 415–424.

Legon, N.W. (1999) A mycological expedition to Puerto Rico. *The Mycologist* 13, 58–62.

Lodge, D.J. (1996a) Fungi of Puerto Rico and the US Virgin Islands: a history of previous surveys, current status, and the future. In: Figueroa Colón, J.C. (ed.) *The Scientific Survey of Puerto Rico and the Virgin Islands.* NY Academy of Sciences, New York, pp. 123–129.

Lodge, D.J. (1996b) Microorganisms. In: Regan, D.P. and Waide, R.B. (eds) *The Food Web of a Tropical Forest.* University of Chicago Press, Chicago, pp. 53–108.

Lodge, D.J. and Pegler, D.N. (1990) The *Hygrophoraceae* of the Luquillo Mountains of Puerto Rico. *Mycological Research* 94, 443–456.

Lodge, D.J., Baroni, T.J., Miller, O.K. Jr and Halling, R.E. (2001) Emerging biogeographic patterns among macrobasidiomycete fungi in the Greater Antilles. In: Zanoni, T. (ed.) *Biogeography of Plants in the Greater Antilles.* New York Botanical Gardens, New York.

Miller, O.K. Jr and Lodge, D.J. (2001) New species of *Amanita* from the Dominican Republic, Greater Antilles. *Mycotaxon* (in press).

Miller, O.K. Jr, Lodge, D.J. and Baroni, T.J. (2000) New and interesting ectomycorrhizal fungi from Puerto Rico, Mona, and Guana Islands. *Mycologia* 92, 558–570.

Moncalvo, J.-M., Lutzoni, F.M., Rehner, S.A., Johnson, J. and Vilgalys, R. (2000) Phylogenetic relationships of agaric fungi based on large subunit ribosomal DNA sequences. *Systematic Biology* 49, 278–305.

Moran, R. and Smith, A. (1999) Phytogeographic relationships between Neotropical and African-Madagascan Pteridophytes. *Proceedings of the XVI International Botanical Congress, 1–7 August 1999, St Louis, Missouri.* Abstracts, addendum, pp. 9–10.

Murrill, W.A. (1910) The *Polyporaceae* of Jamaica. *Mycologia* 2, 183–197.

Murrill, W.A. (1911a) The *Agaricaceae* of Tropical North America. I. *Mycologia* 3, 23–36.

Murrill, W.A. (1911b) The *Agaricaceae* of Tropical North America. II. *Mycologia* 3, 79–91.

Murrill, W.A. (1911c) The *Agaricaceae* of Tropical North America. III. *Mycologia* 3, 189–199.

Murrill, W.A. (1911d) The *Agaricaceae* of Tropical North America. IV. *Mycologia* 3, 271–282.

Murrill, W.A. (1915) *Tropical Polypores.* Published by the author, New York.

Murrill, W.A. (1920a) Light coloured resupinate polypores I. *Mycologia* 12, 77–92.

Murrill, W.A. (1920b) Light coloured resupinate polypores II. *Mycologia* 12, 299–308.

Murrill (1921a) Light coloured resupinate polypores III. *Mycologia* 13, 83–100.

Murrill (1921b) Light coloured resupinate polypores IV. *Mycologia* 13, 171–178.

Nakasone, K., Burdsall, H.H. Jr and Lodge, D.J. (1998) *Phanerochaete flava* in Puerto Rico. *Mycologia* 90, 132–135.

Pegler, D.N. (1983) *Agaric Flora of the Lesser Antilles.* Kew Bulletin Additional Series 9, HMSO, London.

Pegler, D.N. (1987a) A revision of the Agaricales of Cuba 1. Species described by Berkeley & Curtis. *Kew Bulletin* 42, 501–585.

Pegler, D.N. (1987b) A revision of the Agaricales of Cuba: 2. Species described by Earle and Murrill. *Kew Bulletin* 42, 855–888.

Pegler, D.N. (1988) A revision of the Agaricales of Cuba: 3. Keys to families, genera and species. *Kew Bulletin* 43, 53–75.

Pegler, D.N., Lodge, D.J. and Nakasone, K.K. (1998) *Macrocybe* gen. nov (*Tricholomataceae*, tribus *Tricholomatae*). *Mycologia* 90, 494–504.

Poldmaa, K., Samuels, G.J. and Lodge, D.J. (1997) Three new polyporicolous species of *Hypomyces* and their *Cladobotryum* anamorphs. *Sydowia* 49, 80–93.

Rodríguez Gallart, C.A. (1989) Estudios en los macromicetos de la Republica Dominicana. I. *Moscosoa* 5, 142–153.

Rodríguez Gallart, C.A. (1990) Estudios en los macromicetos de la Republica Dominicana. II. *Moscosoa* 6, 202–212.

Rodríguez Gallart, C.A. (1997) Estudios en los macromicetos de la Republica Dominicana. III. *Moscosoa* 9, 145–153.

Ryvarden, L. (1984) Type studies 13. Species described by Berkeley and other authors 1855–1886. *Mycotaxon* 20, 329–363.

Ryvarden, L. (1985) Type studies in the Polyporaceae 17. Species described by W.A. Murrill. *Mycotaxon* 23, 169–198.

Ryvarden, L. (2000a) Studies in neotropical polypores. 2: a preliminary key to neotropical species of *Ganoderma* with a laccate pileus. *Mycologia* 92, 180–191.

Ryvarden, L. (2000b) Studies in Neotropical polypores 5. New and noteworthy species from Puerto Rico and Virgin Islands. *Mycotaxon* 74, 119–129.

Ryvarden, L. (2000c) Studies in neotropical polypores. 7. *Wrightoporia* (Hericiaceae, Basidiomycetes) in tropical America. *Karstenia* 40, 153–158.

Ryvarden, L. (2002) Studies in Neotropical polypores. 8. Poroid fungi from Jamaica – a preliminary check list. *Mycotaxon* (in press).

Schmit, J.P., Murphy, J.F. and Mueller, G.M. (1999) Macrofungal diversity of a temperate oak forest: a test of species richness estimators. *Canadian Journal of Botany* 77, 1014–1027.

Smith, A.H., Theirs, H.D. and Miller, O.K. (1965) The species of *Suillus* and *Fuscoboletinus* of the Priest River Experimental Forest and Vicinity, Priest river, Idaho. *Lloydia* 120–138.

Stevenson, J.A. (1970) *Fungi of Puerto Rico and the American Virgin Islands.* Publication 23, Reed Herbarium, Baltimore, Maryland.

Swartz, O.P. (1788) *Nova genera et species plantarum.* Holmiae, Stockholm, Sweden.

Tropical Brown- and Black-spored Mexican Agarics with Particular Reference to *Gymnopilus* 5

L. Guzmán-Dávalos

Departamento de Botánica y Zoología, Universidad de Guadalajara, Apdo postal 1-139, Zapopan, Jalisco 45110, Mexico

Introduction

There are relatively few published studies on tropical Mexican fungi, and dark-spored agarics appear to be a particularly neglected group. Mexican surveys have been directed mainly to the temperate forests and to light-spored genera, although modern mycology in Mexico started with the dark-spored, hallucinogenic mushrooms (Guzmán, 1990). Many mycologists preferred to collect in *Pinus–Quercus* forests where the conditions are more favourable than in the tropical regions. As Nishida (1989) pointed out, 'A mushroom collector there (warm tropical climates) must gather his fungi early in the morning, take his notes, and dry the specimens in a timely fashion'. Maintenance and transport of dry mushrooms in the tropics are also a problem; high atmospheric humidity reaching the specimens allows rapid growth of moulds.

To address the tropical dark-spored (brown and black) Mexican agarics, this chapter considers the mycodiversity of Mexico and the vegetation types present in the country with emphasis on the tropics. The history of Mexican studies of agarics and how they have contributed to the current state of knowledge is then briefly mentioned. Finally, the special features of *Gymnopilus* are described.

Mexican Biodiversity and Vegetation Types

Mexico has a privileged position in North America from a biodiversity point of view. Situated at the transition between the Neotropical and Nearctic realms,

 Tropical Mycology, Vol. 1, *Macromycetes* (eds R. Watling, J.C. Frankland, A.M. Ainsworth, S. Isaac and C.H. Robinson)

Mexico has representatives from both, as well as endemic species owing to the confluence of the two regions. For this reason, Mexico is considered a country of megadiversity (Mittermeier, 1988). It presents a mosaic of vegetation types, with temperate forests (i.e. *Quercus* and conifer forests) and alpine vegetation in the mountains; xerophytic vegetation and grasslands on the Mexican Plateau; tropical and subtropical vegetation on the coasts and in the south of the country and aquatic vegetation represented in some regions. The tropical and subtropical vegetation in Mexico is represented by tropical rainforests (or tropical evergreen forests), tropical subdeciduous and deciduous forests and the mesophytic or cloud forests (Rzedowski, 1978). The last have also been included in 'the humid temperate zone in areas with subtropical climates' (Toledo and Ordóñez, 1993).

Tropical and subtropical vegetation covers less than 30% of Mexico (Neyra and Durand, 1998). Tropical rainforest is present only in southern Mexico, mainly in Chiapas. This forest is very humid, with fungi fruiting throughout the year. The majority of these fungi are lignicolous, and few are humicolous. High temperatures and humidity favour rapid decomposition of the humus, so the substratum of humicolous fungi is available for only short periods of time (Guzmán-Dávalos and Guzmán, 1979). In contrast, the subdeciduous and deciduous tropical forests are dry in winter and spring and not as rich in fungi. They have sufficient humidity to support fungal fruiting during only a few summer months.

The mesophytic forest is a type of subtropical humid vegetation. It is also called cloud forest owing to the high relative humidity, which creates a mist for most of the year. This condition promotes the existence of a very rich mycobiota, which has been well studied in Mexico relative to other tropical vegetation types. According to Guzmán (1996), the mesophytic forest has many similarities with the deciduous forests of eastern USA, and they have many fungi in common. In Mexico, the mesophytic forest covers only 0.07% of the land surface and it is considered to be one of the most threatened ecosystems (Flores-Villela and Gerez, 1994).

Mycological Studies on Dark-spored Agarics

Studies of Mexican agarics began with Murrill in 1910, although there are some isolated records from the 18th century (Guzmán-Dávalos, 2000). Murrill described many dark-spored agarics from tropical regions of Mexico, including 63 agarics of which 25 were dark-spored and belonged to the genera: *Agaricus, Agrocybe, Bolbitius, Coprinus, Cortinarius, Galerina, Inocybe, Melanotus, Phaeomarasmius, Pholiota, Psathyrella* and *Simocybe* (Halling, 1986). The start of modern investigations of Mexican agarics was directly related to the rediscovery of hallucinogenic mushrooms in the 1950s, with the works of Wasson and Heim (e.g. Heim, 1956; Heim and Wasson, 1958) followed by Singer

(1957, 1958a,b,c; Singer and Smith, 1958a,b). Singer's first visit to Mexico was motivated by the curiosity surrounding these mushrooms. From there, his interest developed into a study of Mexican agarics, and he played an outstanding role in the study and understanding of Mexican *Agaricales*, especially of those from the tropics, and was a great influence in Mexican mycology. His authority was so great that Guzmán (1994) considered that fungal taxonomy in Mexico could be historically divided between pre- and post-Singer. Also during the 1950s, Guzmán initiated his research with Mexican mushrooms, specifically with *Psilocybe* (Guzmán, 1958, 1959), under the great influence of Singer. Guzmán has worked with this genus ever since and it is no coincidence that *Psilocybe* is now the most studied fungal genus in Mexico.

Most of the Mexican fungal inventories include few species of dark-spored agarics, although there are some exceptions. Guzmán and Johnson (1974) considered 32 species of basidiomycetes collected in a tropical zone in Chiapas, of which 11 were dark-spored agarics. Welden and Guzmán (1978) and Welden *et al.* (1979) listed more than 400 species, including about 50 dark-spored agarics of temperate, subtropical and tropical vegetation of south-western Mexico. In the same region, Guzmán-Dávalos and Guzmán (1979) found 170 species, including 21 dark-spored agarics, but only ten of them were from tropical regions. Guzmán (1983b) reported 274 species from the Yucatan Peninsula, of which 25 were dark-spored tropical agarics, and of these eight were *Psathyrella* species. Frutis and Guzmán (1983) made an inventory of the macroscopic fungi of Hidalgo State recording 422 species including 62 with dark spores. This did not include tropical localities, only mesophytic forest with 24 species of dark-spored agarics, mainly of *Psilocybe* and *Inocybe*. Guzmán and Guzmán-Dávalos (1984) recorded four dark-spored agarics among 35 species found in Veracruz State. From a list of 270 species of higher fungi from Michoacán State, 26 were dark-spored agaric taxa but only eight were from subtropical (mainly mesophytic forest) or tropical regions (Díaz-Barriga *et al.*, 1988).

The genera of dark-spored agarics that have currently received most attention in Mexico are *Agaricus*, *Coprinus*, *Cortinarius* (only subgenus *Dermocybe*), *Gymnopilus*, *Hypholoma*, *Inocybe*, *Micropsalliota*, *Panaeolus*, *Phaeocollybia*, *Psathyrella* and *Psilocybe*. Recently, *Crepidotus* has also been studied, for example, Bandala *et al.* (1999) studied two species of Mexican *Crepidotus*, one from temperate and mesophytic forests and the other from tropical rainforest. However, if only the tropical genera, or those with tropical representatives, had been considered, we would have a very limited impression of the tropical dark-spored Mexican agarics.

The genus *Agaricus* was studied by Gutiérrez-Ruiz and Cifuentes (1990). They considered 14 species from the subgenus *Agaricus*, from mesophytic forest and *Pinus–Quercus* forest. *Coprinus* is poorly known in Mexico, with very few descriptions in the literature (e.g. Pérez-Silva, 1976). There is only one publication concerning *Cortinarius* subgenus *Dermocybe*, and the six species considered are from temperate forests (Sánchez Macías *et al.*, 1987). *Hypholoma* has been studied by Guzmán as part of his research on genera related to *Psilocybe* (Guzmán, 1975, 1999), and six species are known from the country. *Inocybe* was studied by Pérez-Silva (1967, 1976–1982) who recorded about 60 species for Mexico, mainly from temperate regions. An unusual tropical species was described by Guzmán (1982), *Inocybe tropicalis* Guzmán, from Yucatan.

Micropsalliota, a tropical and subtropical genus mainly from Asia, has been studied by Heinemann. In Mexico, three species are known, two of them endemic and from a mesophytic forest (Guzmán-Dávalos, 1992; Guzmán-Dávalos and Heinemann, 1994). There are certainly more species waiting to be described in collections already made from tropical forests by the present author and collaborators. In Guzmán and Pérez-Patraca's monograph (1972) of the Mexican species of *Panaeolus s. l.*, 11 taxa, 9 of them from tropical regions, were considered. Among them, *Panaeolus antillarum* (Fr.) Dennis and *Copelandia cyanescens* (Berk. & Broome) Singer are the most common in the country.

Mexican species of *Phaeocollybia* have been studied extensively by Bandala and collaborators. To date 19 species are known, mainly from temperate forests, although some are from mesophytic forests (Guzmán *et al.*, 1987; Bandala *et al.*, 1989, 1996; Bandala and Montoya, 1994). Nineteen is high compared with the 45 species known throughout the world, according to Hawksworth *et al.* (1995). Tropical species of *Psathyrella* in Mexico have been studied by Guzmán and Johnson (1974), Guzmán (1983b) and Guzmán *et al.* (1988). For example, *Psathyrella smithii* Guzmán was described from a tropical rainforest in Chiapas and *Psathyrella asperospora* (Cleland) Guzmán, Bandala & Montoya (= *Lacrymaria asperospora* (Cleland) Watling, = *Psathyrella sepulchralis* Singer, Smith & Guzmán) is common in disturbed areas of the cloud forest.

As mentioned above, *Psilocybe* is the most studied genus in Mexico, not only among the dark-spored agarics but also of all the agarics. In his monograph of *Psilocybe* and the supplement published in 1995, Guzmán (1983a, 1995) accepted 172 taxa of which 40 are known from Mexico. A very common species in tropical and subtropical regions is *Psilocybe cubensis* (Earle) Singer, and because of its size, abundance and ease of cultivation it is the most commonly used hallucinogenic mushroom, albeit illegally. *Psilocybe uxpanapensis* Guzmán is one of the few hallucinogenic fungi that grows only in tropical rainforests, but it is in danger of extinction because of rapid deforestation. New tropical species remain to be described. Only recently Guzmán described two species from a mesophytic forest in Jalisco (Guzmán, 1998) and three from Veracruz, two from mesophytic forest and one from alpine forest at 3300 m (Guzmán *et al.*, 1999).

The Genus *Gymnopilus*

This genus belongs to *Cortinariaceae*, *Cortinarieae* (Singer, 1986). It is distinguished macroscopically by the yellow to ferruginous basidiomata and lamellae. Microscopically, the spores are ellipsoid, or sometimes widely ellipsoid, verrucose and ferruginous. The cystidia are an important taxonomic feature; in many species they are lageniform, capitate or subcapitate. An important ecological feature of the genus is its lignicolous habitat. *Gymnopilus* can grow on wood in different states of decay, including very decayed wood. The only monograph on the genus is that of Hesler (1969), in which only one species from Mexico was considered.

Currently, 31 species of the genus are known in Mexico, among 175 known in the world (Hawksworth *et al.*, 1995). It is interesting to note that of the 31 species, 17 species are unique to Mexico. Six are found only in the USA and Mexico. Four, among them *Gymnopilus penetrans* (Fr.: Fr.) Murrill and *Gymnopilus sapineus* (Fr.) R. Maire, have a worldwide distribution. Two, *Gymnopilus robustus* and *Gymnopilus rugulosus*, grow in Mexico and Central America. *Gymnopilus palmicola* is in Mexico, USA and Cuba, and *Gymnopilus lateritius* in Mexico and the Lesser Antilles (Guzmán-Dávalos, 1993; Guzmán-Dávalos and Guzmán, 1995; Guzmán-Dávalos and Ovrebo, 2001) (Fig. 5.1).

Of the 31 species known in Mexico, 58% are from temperate forests consisting chiefly of *Quercus* and *Pinus*; 42% from the warm regions: comprising seven species from the tropics, two from mesophytic forest and four that can be found in both tropical and temperate regions with some tropical influence (Fig. 5.2). These numbers reflect the status of *Gymnopilus* in the tropics, a genus which used to be cited mainly from temperate regions (Hesler, 1969; Høiland, 1990). Until recently, Mexican mycological expeditions concentrated on temperate and subtropical forests, leaving the tropical regions without due attention. The number of tropical species is likely to increase as these regions are better explored. This is also happening in other tropical regions, for example, Høiland (1998) recently described two new species from Africa.

The tropical, subtropical and temperate species are shown in Table 5.1. One of the largest basidioma of the genus is produced by *G. robustus*, described from Mexico and recently collected in Central America (Costa Rica and Panama) (Guzmán-Dávalos, 1994; Guzmán-Dávalos and Ovrebo, 2000). It fruits only on

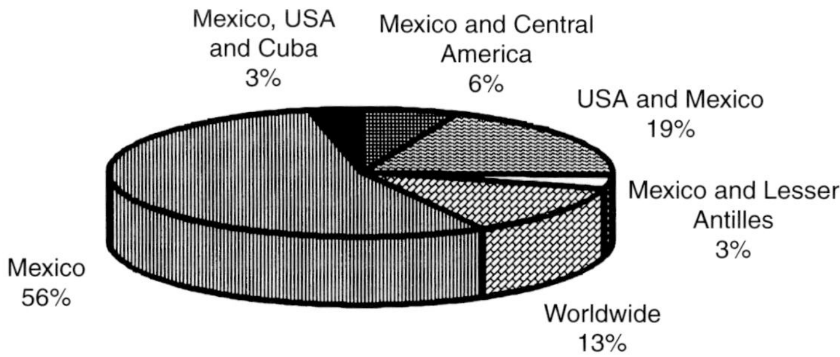

Fig. 5.1. World distribution of the Mexican species of *Gymnopilus*.

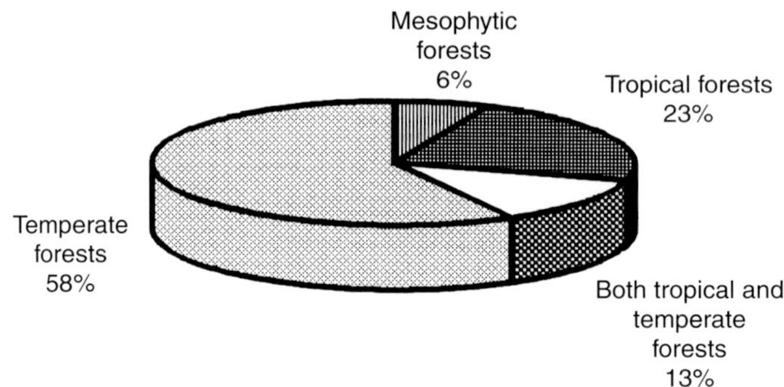

Fig. 5.2. Mexican species of *Gymnopilus* according to vegetation type.

stems or adventitious root masses of palms in tropical regions, but may have an even wider Neotropical distribution. *G. robustus* is easy to identify microscopically because the spores are widely ellipsoid to subglobose and have a conspicuous perisporium. *Gymnopilus tuxtlense* is known only from tropical regions of Mexico (Guzmán-Dávalos, 1994). It is easy to recognize owing to its small and slender basidiomata, papillate pileus and a yellow pubescent stipe base. Another producer of large basidiomata is *G. rugulosus*. It was described from Mexico (Valenzuela *et al.*, 1981), and recently was recorded in Costa Rica (Guzmán-Dávalos and Ovrebo, 2000). In both countries it is abundant in subtropical forests. Owing to its tuberculate, widely ellipsoid to subglobose spores, *G. rugulosus* is also very easy to identify under the microscope. *Gymnopilus lepidotus* is known from subtropical and tropical rainforest of Florida, USA and Mexico (Hesler, 1969; Guzmán-Dávalos, 1996). It has small basidiomata with tiny erect purplish squamules. *G. palmicola* was described from Cuba (Murrill, 1913). The only colour photograph in Singer's book (Singer, 1986), as *Gymnopilus aculeatus* (Bres. & Roumg.) Singer, is of this species. The specimen in the photograph was found in a greenhouse in the USA, growing in a living orchid collected from Oaxaca in Mexico. It is also known from Florida, USA (Guzmán-Dávalos and Guzmán, 1995). *Gymnopilus subpurpuratus* is known both from *Pinus* and cloud forests only in Mexico. It is probably a hallucinogenic species since its basidiomata stain blue-green when bruised (Guzmán-Dávalos and Guzmán, 1991).

The destruction of habitats, especially tropical ones, is alarming. We need to place special emphasis on the study of tropical regions, otherwise it will be too late. We know that there are still many tropical species of *Gymnopilus* waiting to be named, some of them already collected but not yet studied and many others not even collected. It is almost certain that tropical species common in other countries, such as *Gymnopilus chrysopellus* (Berk. & M. A. Curtis) Murrill (Hesler, 1969; Kreisel, 1971) or *Gymnopilus dilepis* (Berk. & Broome) Singer (Pegler, 1986) will be present in Mexico. Also there are many other dark-spored agarics not described from the Mexican tropics. If in well-studied genera, such

Table 5.1. Tropical species of *Gymnopilus* from Mexico, according to vegetation type.

Tropical forests (evergreen, deciduous and subdeciduous)
G. hemipenetrans Guzmán-Dávalos
G. lateritius (Pat.) Murrill
G. medius Guzmán-Dávalos
G. robustus Guzmán-Dávalos
G. cf. *subdryophilus* Murrill
G. subearlei Valenz., Guzmán & J. Castillo
G. tuxtlense Guzmán-Dávalos
Subtropical forests (mesophytic)
G. rugulosus Valenz., Guzmán & J. Castillo
G. subgeminellus Guzmán-Dávalos & Guzmán
Tropical, subtropical and temperate forests (in many cases the temperate forests have tropical influence)
G. fulvosquamulosus Hesler
G. lepidotus Hesler
G. palmicola Murrill
G. subpurpuratus Guzmán-Dávalos & Guzmán

as *Psilocybe*, there are still many unnamed species, what finds can be expected in genera not yet studied? Genera such as *Conocybe*, *Cystoagaricus*, *Melanotus*, *Phaeomarasmius*, *Pyrrhoglossum*, among others, need to be addressed.

Acknowledgements

My thanks to Roy Watling and Tony Whalley of the British Mycological Society Millennium Meeting for inviting me to address the symposium. Financial support was provided by the British Mycological Society, CONABIO (National Commission for the Study and Use of Biodiversity of Mexico), Idea Wild, SIMORELOS-CONACyT (National Council of Science and Technology of Mexico) and University of Guadalajara. I thank also Gastón Guzmán for his critical review of the paper. Pablo Munguía kindly helped with the English translation, and the editors of the Tropical Mycology volumes contributed in making the English more concise.

References

Bandala, V.M. and Montoya, L. (1994) Further investigations on *Phaeocollybia* with notes on infrageneric classification. *Mycotaxon* 52, 397–422.

Bandala, V.M., Guzmán, G. and Montoya, L. (1989) Additions to the knowledge of *Phaeocollybia* (*Agaricales*, *Cortinariaceae*) in Mexico with description of new species. *Mycotaxon* 35, 127–152.

Bandala, V.M., Montoya, L., Guzmán, G. and Horak, E. (1996) Four new species of *Phaeocollybia*. *Mycological Research* 100, 239–243.

Bandala, V.M., Montoya, L. and Moreno, G. (1999) Two *Crepidotus* from Mexico with notes on selected type collections. *Mycotaxon* 72, 403–416.

Díaz-Barriga, H., Guevara-Fefer, F. and Valenzuela, R. (1988) Contribución al conocimiento de los macromicetos del Estado de Michoacán. *Acta Botánica Mexicana* 2, 21–44.

Flores-Villela, O. and Gerez, P. (1994) *Biodiversidad y conservación en México: vertebrados, vegetación y uso de suelo*. CONABIO and UNAM, Mexico, DF, Mexico.

Frutis, I. and Guzmán, G. (1983) Contribución al conocimiento de los hongos del Estado de Hidalgo. *Boletín de la Sociedad Mexicana de Micología* 18, 219–265.

Gutiérrez-Ruiz, J. and Cifuentes, J. (1990) Contribución al conocimiento del género *Agaricus* subgénero *Agaricus* en México, I. *Revista Mexicana de Micología* 6, 151–177.

Guzmán, G. (1958) El hábitat de *Psilocybe muliercula* Singer & Smith (= *Ps. wassonii* Heim), agaricáceo alucinógeno mexicano. *Revista de la Sociedad Mexicana de Historia Natural* 19, 215–229.

Guzmán, G. (1959) Sinopsis de los conocimientos sobre los hongos alucinógenos mexicanos. *Boletín de la Sociedad Botánica de México* 24, 14–34.

Guzmán, G. (1975) New and interesting species of *Agaricales* of Mexico. In: Bigelow, H. and Thiers, H. (eds) *Studies on Higher Fungi. Beihefte zur Nova Hedwigia* 51. Cramer, Vaduz, Lichtenstein, pp. 99–118.

Guzmán, G. (1982) New species of fungi from the Yucatan Peninsula. *Mycotaxon* 16, 249–261.

Guzmán, G. (1983a) The genus *Psilocybe*. *Beihefte zur Nova Hedwigia* 74. Cramer, Vaduz, Lichtenstein.

Guzmán, G. (1983b) Los hongos de la Península de Yucatán. II. Nuevas exploraciones y adiciones micológicas. *Biotica* 8, 71–100.

Guzmán, G. (1990) La micología en México. *Revista Mexicana de Micología* 6, 11–28.

Guzmán, G. (1994) *In memoriam* Dr. Rolf Singer 1907–1994. *Revista Mexicana de Micología* 10, 199–201.

Guzmán, G. (1995) Supplement to the monograph of the genus *Psilocybe*. In: Petrini, O. and Horak, E. (eds) *Taxonomic monographs of Agaricales. Bibliotheca Mycologica* 159. Cramer, Berlin, pp. 91–141.

Guzmán, G. (1996) Observations on some fungi from Louisiana and Mississippi in comparison with those of Mexico. *Tulane Studies in Zoology and Botany* 30, 69–74.

Guzmán, G. (1998) Las especies de *Psilocybe* (*Fungi, Basidiomycotina, Agaricales*) conocidas de Jalisco (México) y descripción de dos nuevas para la ciencia. *Acta Botánica Mexicana* 43, 23–32.

Guzmán, G. (1999) New combinations in *Hypholoma* and information on the distribution and properties of the species. *Documents Mycologiques* 29(114), 65–66.

Guzmán, G. and Guzmán-Dávalos, L. (1984) Nuevos registros de hongos en el Estado de Veracruz. *Boletín de la Sociedad Mexicana de Micología* 19, 221–244.

Guzmán, G. and Johnson, P.D. (1974) Registros y especies nuevas de los hongos de Palenque, Chiapas. *Boletín de la Sociedad Mexicana de Micología* 8, 73–105.

Guzmán, G. and Pérez-Patraca, A.M. (1972) Las especies conocidas del género *Panaeolus* en México. *Boletín de la Sociedad Mexicana de Micología* 6, 17–53.

Guzmán, G., Bandala, V.M. and Montoya, L. (1987) The known species of *Phaeocollybia* (*Agaricales, Cortinariaceae*) in Mexico. *Mycotaxon* 30, 221–238.

Guzmán, G., Montoya-Bello, L. and Bandala-Muñoz, V.M. (1988) A new species of *Psathyrella* (*Agaricales, Coprinaceae*) from Mexico with discussions on the known species. *Brittonia* 40, 229–234.
Guzmán, G., Ramírez-Guillén, F., Tapia, F. and Navarro, P. (1999) Las especies del género *Psilocybe* (*Fungi, Basidiomycotina, Agaricales*) conocidas de Veracruz (México). *Acta Botánica Mexicana* 49, 35–46.
Guzmán-Dávalos, L. (1992) First record of the genus *Micropsalliota* (*Agaricaceae, Agaricales*) in Mexico. *Mycotaxon* 43, 199–205.
Guzmán-Dávalos, L. (1993) Contribución al conocimiento del género *Gymnopilus* (*Agaricales, Cortinariaceae*) en México. MSc thesis, Facultad de Ciencias, UNAM, Mexico, DF, Mexico.
Guzmán-Dávalos, L. (1994) New species of *Gymnopilus* (*Agaricales, Cortinariaceae*) from Mexico. *Mycotaxon* 50, 333–348.
Guzmán-Dávalos, L. (1996) New records of the genus *Gymnopilus* (*Agaricales, Cortinariaceae*) from Mexico. *Mycotaxon* 59, 61–78.
Guzmán-Dávalos, L. (2000) Los estudios sobre Agaricales en México. *Boletín IBUG (Instituto de Botánica, Universidad de Guadalajara)* 6, 279–295.
Guzmán-Dávalos, L. and Guzmán, G. (1979) Estudio ecológico comparativo entre los hongos (macromicetos) de los bosques tropicales y los de coníferas del Sureste de México. *Boletín de la Sociedad Mexicana de Micología* 13, 89–125.
Guzmán-Dávalos, L. and Guzmán, G. (1991) Additions to the genus *Gymnopilus* (*Agaricales, Cortinariaceae*) from Mexico. *Mycotaxon* 41, 43–56.
Guzmán-Dávalos, L. and Guzmán, G. (1995) Toward a monograph of the genus *Gymnopilus* (*Cortinariaceae*) in Mexico. *Documents Mycologiques* 25(98–100), 197–212.
Guzmán-Dávalos, L. and Heinemann, P. (1994) *Micropsalliota subalpina* nov. sp. (*Agaricaceae*) from Mexico. *Bulletin Jardin Botanique National de Belgique* 63, 195–199.
Guzmán-Dávalos, L. and Ovrebo, C. (2001) Some species of *Gymnopilus* from Costa Rica and Panama. *Mycologia* 93, 398–404.
Halling, R.E. (1986) An annotated index to species and infraspecific taxa of Agaricales and Boletales described by William A. Murrill. *Memoirs of the New York Botanical Garden* 40, 1–120.
Hawksworth, D.L., Kirk, P.M., Sutton, B.C. and Pegler, D.N. (1995) *Ainsworth & Bisby's Dictionary of the Fungi*, 8th edn. CAB International, Wallingford, UK.
Heim, R. (1956) Les champignons divinatoires utilisés dans les rites des Indiens mazatèques, recueillis au cours de leur premier voyage au Mexique, en 1953, par Mme. Valentina Pavlovna Wasson et M.R. Gordon Wasson. *Comptes Rendus des Séances de l'Académie des Séances* 242, 965–968.
Heim, R. and Wasson, R.G. (1958) *Les champignons hallucinogènes du Mexique*. Archives du Muséum National d'Histoire Naturelle 6, 1–324.
Hesler, L.R. (1969) *North American Species of Gymnopilus. Mycological Memoirs* 3. Hafner, New York.
Høiland, K. (1990) The genus *Gymnopilus* in Norway. *Mycotaxon* 39, 257–279.
Høiland, K. (1998) *Gymnopilus purpureosquamulosus* and *G. ochraceus* spp. nov. (*Agaricales, Basidiomycota*) – two new species from Zimbabwe. *Mycotaxon* 69, 81–85.
Kreisel, H. (1971) *Clave para la identificación de los macromicetos de Cuba*, 4. Ciencias Biológicas No. 16, Centro de Información Científica y Técnica, Universidad de La Habana, La Habana, Cuba.

Mittermeier, R.A. (1988) Primate diversity and the tropical forest: case studies from Brazil and Madagascar and the importance of the megadiversity countries. In: Wilson, E. (ed.) *Biodiversity*. National Academic Press, Washington, DC, pp. 145–154.

Murrill, W.A. (1910) Agaricales, *Boletaceae*, *Agaricaceae* (pars) and *Cantharelleae*, *Lactarieae* (pars). In: *North American Flora* 9 (3). New York Botanical Garden, New York, pp. 133–200.

Murrill, W.A. (1913) The *Agaricaceae* of tropical North America, VI. *Mycologia* 5, 18–36.

Neyra, L. and Durand, L. (1998) Biodiversidad. In: CONABIO. *La diversidad biológica de México: estudio de país, 1998*. Comisión Nacional para el Conocimiento y Uso de la Biodiversidad, Mexico, DF, Mexico, pp. 61–102.

Nishida, F.H. (1989) Review of mycological studies in the Neotropics. In: Campbell, D. and Hammond, D.H. (eds) *Floristic Inventory of Tropical Countries*. New York Botanical Garden, New York, pp. 494–533.

Pegler, D.N. (1986) *Agaric Flora of Sri Lanka*. Kew Bulletin Additional Series XII. HMSO, London.

Pérez-Silva, E. (1967) Les Inocybes du Mexique. *Anales del Instituto de Biología, Universidad Nacional Autónoma de México* 38, ser. Bot., 1–60.

Pérez-Silva, E. (1976) Hongos fimícolas de México, III. Especies poco conocidas del género *Coprinus* (*Agaricales*). *Boletín de la Sociedad Mexicana de Micología* 10, 23–26.

Pérez-Silva, E. (1976–1982) Nuevos registros en México para algunas especies del género *Inocybe* (Agaricales). *Anales del Instituto de Biología, Universidad Nacional Autónoma de México* 47–53, ser. Bot., 177–188.

Rzedowski, J. (1978) *Vegetación de México*. Editorial Limusa, Mexico, DF, Mexico.

Sánchez Macías, E., Pérez-Silva, E. and Pérez Amador, C. (1987) Consideraciones quimiotaxonómicas para el estudio de algunas especies del género *Dermocybe* (Cortinariaceae) en México. *Revista Mexicana de Micología* 3, 189–202.

Singer, R. (1957) Fungi mexicani, series prima – Agaricales. *Sydowia* 11, 354–374.

Singer, R. (1958a) Observations on agarics causing cerebral mycetism. *Mycopathologia et Mycologia Applicada* 9, 261–284.

Singer, R. (1958b) Mycological investigation on Teonanacatl, the Mexican hallucinogenic mushroom. Part I. The history of Teonanacatl, field work and culture work. *Mycologia* 50, 239–261.

Singer, R. (1958c) Fungi Mexicani, series secunda – Agaricales. *Sydowia* 12, 221–243.

Singer, R. (1986) *The Agaricales in Modern Taxonomy*. Koeltz, Koenigstein, Germany.

Singer, R. and Smith, A.H. (1958a) Mycological investigation on Teonanacatl, the Mexican hallucinogenic mushroom. Part II. A taxonomic monograph of *Psilocybe*, section *Caerulescentes*. *Mycologia* 50, 262–303.

Singer, R. and Smith, A.H. (1958b) New species of *Psilocybe*, Section *Caerulescentes*. *Mycologia* 50, 141–142.

Toledo, V.M. and Ordoñez, M.J. (1993) The biodiversity scenario of Mexico: a review of terrestrial habitats. In: Ramamoorthy, T.P., Bye, R., Lot, A. and Fa, J. (eds). *Biological Diversity of Mexico. Origins and Distribution*. Oxford University Press, New York, pp. 757–778.

Valenzuela, R., Guzmán, G. and Castillo, J. (1981) Descripciones de especies de macromicetos poco conocidas en Mexico, con discusiones sobre su ecología y distribución. *Boletín de la Sociedad Mexicana de Micología* 15, 67–120.

Welden, A. and Guzmán, G. (1978) Lista preliminar de los hongos, líquenes y mixomicetos de las regiones de Uxpanapa, Coatzacolacos, Los Tuxtlas, Papaloapan y Xalapa (parte de los estados de Veracruz y Oaxaca). *Boletín de la Sociedad Mexicana de Micología* 12, 59–102.

Welden, A.L., Dávalos, L. and Guzmán, G. (1979) Segunda lista de los hongos, líquenes y mixomicetos de las regiones de Uxpanapa, Coatzacolacos, Los Tuxtlas, Papaloapan y Xalapa (México). *Boletín de la Sociedad Mexicana de Micología* 13, 151-161.

Patterns of Polypore Distribution in the Lesser Sunda Islands, Indonesia. Is Wallace's Line Significant?

6

M. Núñez,[1] Suhirman[2] and J.N. Stokland[3]

[1]*Mycoteam, Forskningsveien 3b, PO Box 5, Blindern, N-0313 Oslo, Norway;* [2]*Indonesian Botanic Gardens, Jl. Juanda 13 Bogor 16122, Indonesia;* [3]*Norwegian Institute of Land Inventory, PO Box 115, N-1430 Ås, Norway*

Introduction

Indonesia has often played an important role in biogeographical research owing to its key position at the meeting point of the Asian and the Australian tectonic plates (Audley-Charles, 1987). Based on distributional patterns of mammals, birds and butterflies, Wallace (1860) drew a line across the Indonesian archipelago to mark the borderline between the Asian and the Australian biotas. Today, the biological significance of Wallace's line is considered outdated for many groups of organisms (Simpson, 1977; van Steenis, 1979; Touw, 1992; Hisheh *et al.*, 1998). However, some organisms have a distribution limited either to the Asian or to the Australian side of Wallace's line, as in the case for most genera of conifers and taxads (Schuster, 1972).

The Lesser Sunda Islands consist of two geologically distinct archipelagos. The northern archipelago, including Bali, Lombok, Sumbawa, Flores and Wetar, is volcanic in origin. It was formed during the Pliocene after the collision between the Australian and the Asian plate, about 15×10^6 years ago, Wallace's line lying between Bali and Lombok (Fig. 6.1). The islands of the southern archipelago, including Sumba, Timor and Babar, are non-volcanic and appear to belong to the Australian plate (Audley-Charles, 1987; Veevers, 1991). Little mycological research has been done in the Lesser Sunda Islands and no publications documenting their polypore mycota were found during this study, but Pirozynski (1983) and Sims *et al.* (1997) have discussed several distributional patterns of Australasian fungi.

This chapter addresses the significance of Wallace's line for polypore distribution in some of the Lesser Sunda Islands from the northern archipelago, namely Bali, Lombok and Sumbawa. These islands were chosen for this study because of the position of the Wallace line coinciding as it does with the geological boundary between the Asian and Australian plates (Musser, 1981). When these plates approached each other, two biotas of independent origin were physically mixed.

In addition, the proximity of these islands to Bali Botanical Garden made logistics easier and Bali, Lombok and Sumbawa lie aligned to the east with Sumatra and Java, which have a relatively well known mycota of polypores (Suhirman and Núñez, 1995, 1996).

Similarities in polypore species composition between the different collecting localities were quantified. Using literature records, this analysis was extended to include Sumatra, Java, Papua New Guinea (PNG) and Australia in an attempt to explain polypore distribution patterns in Australasia according to Croizat's (1958) model of generalized tracks.

Polypore species lists were compiled for Sumatra, Java, Bali, Lombok, Sumbawa, PNG and Australia. These lists are based on published records (Imazeki, 1952; Cunningham, 1965; Steyaert, 1972; Ryvarden, 1981; Quanten, 1993; Suhirman and Núñez, 1995, 1996, 1998), and the authors' collections from Java, Bali, Lombok and Sumbawa at the beginning of the rainy season during 1994 and 1996. The type of forest in each collecting locality was specified (Table 6.1).

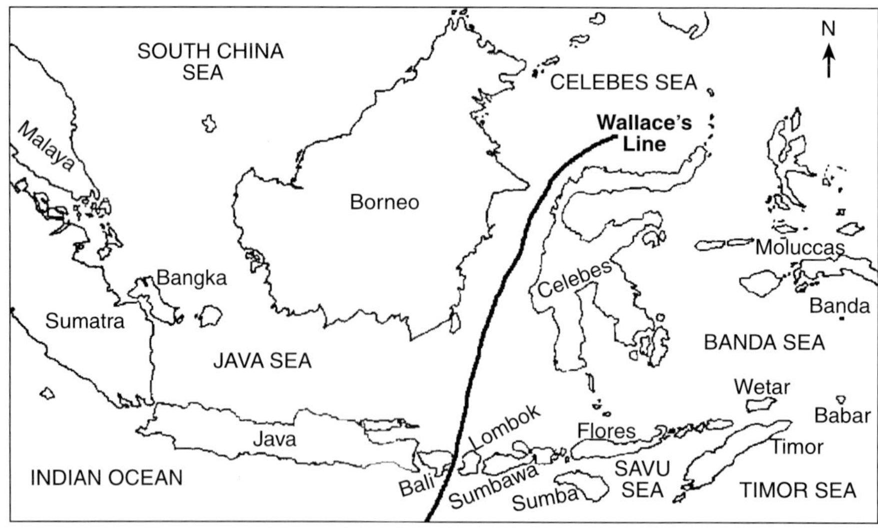

Fig. 6.1. Wallace's line lying between Bali and Lombok.

Table 6.1. Collecting localities in the Lesser Sunda Islands.

Bali monsoon
Bali: Teluk Terima, Bali-Barat National Park, Jembrana province, 0–250 m a.s.l., 20–21 May 1994. Monsoon forest.

Bali lowland
Bali: Gunnung Klatakan, Bali-Barat National Park, Jembrana province, 500–690 m a.s.l., 22–23 May 1994. Lowland rainforest.

Bali cloud
Bali: Bukit Tapak, 1300–1700 m a.s.l., 25–26 May 1994. Cloud forest with *Casuarina junghuhniana* and *Podocarpus* sp.

Lombok lowland
Lombok: Torean, Rinjani Mountain, 700–900 m a.s.l., 28–29 May 1994. Lowland rainforest.

Lombok cloud
Lombok: Ponsuk forest, Sembalun, 1500–1600 m a.s.l., 1 June 1994 and 19–20 October 1996. Cloud forest.

Sumbawa monsoon
Sumbawa: Kawinda, Tambora Mt, 450–860 m a.s.l., 25 October 1996. Monsoon forest.

Sumbawa cloud
Sumbawa: Semonkat Brombosang, 1200 m a.s.l., 23–24 October 1996. Cloud forest.

The similarity index $S = 2c/(a + b)$ was used to quantify similarities in species composition between the different localities. In this formula, *a* and *b* are the number of taxa in locality *a* and *b* respectively, and *c* is the number of shared taxa between the two localities. The value of the index varies between 0 (no shared taxa) and 1.0 (identical species composition in locality *a* and *b*).

Results

A total of 102 species were identified from 626 collections sampled in Bali, Lombok and Sumbawa as listed in Table 6.2. However, 15 collections of 6 or 7 species remain unidentified resulting from either nomenclatural or taxonomic problems. In Java, Sumatra, Australia, PNG and the Lesser Sunda Islands the total was 375. *Ceriporia ferruginicincta* (Murrill) Ryv., and *Skeletocutis diluta* (Rajchenb.) A. David & Rajchenb. are new records for Asia.

The world distribution of the species collected in Bali, Lombok and Sumbawa is shown in Table 6.3. Most of the species were either pantropical (52%) or palaeotropical (25%). Australian species have not yet been recorded from the area, and only two Australasian species were collected, *Elmerina cladophora* (Berk.) Bres. and *Oxyporus cervino-gilvus* (Junghuhn) Ryv., both of which were found in the cloud forests of Lombok and Sumbawa. Two holarctic species were found: *Antrodiella semisupina* (Berk. & M.A. Curtis) Ryv. in Lombok cloud forest, and *Oxyporus populinus* (Fr.) Donk in Sumbawa cloud forest.

Table 6.2. Polypore species collected in Bali, Lombok and Sumbawa forest types.

Species	Bali monsoon	Bali lowland	Bali cloud	Lombok lowland	Lombok cloud	Sumbawa monsoon	Sumbawa cloud
Antrodiella liebmannii		*	*	*	*		
A. semisupina					*		
Bjerkandera adusta				*	*	*	
Ceriporia ferruginicincta						*	
C. mellea						*	
C. xylostromatoides							*
Cerrena meyenii						*	
Chaetoporellus latitans					*		*
Coriolopsis asper	*		*	*		*	
C. brunneo-leuca						*	
C. byrsina						*	*
C. caperata	*						
C. floccosa	*	*		*		*	*
C. glabrorigens							
C. sanguinaria		*					
C. telfarii	*						
Cyclomyces fuscus		*	*				
C. setiporus		*	*	*	*		
C. tabacinus			*	*	*		
Dictyopanus pusillus			*		*		
Earliella scabrosa	*	*	*		*		*
Echinochaete brachyporus		*					*
E. ruficeps	*						
Echinoporia hydnophora					*		
Elmerina cladophora					*	*	
Flavodon flavus	*	*		*	*	*	*

Fomitopsis carnea		*		*			
F. feei						*	*
F. rhodophaea	*	*	*				
G. applanatum			*		*		
G. australe	*						*
G. colossum	*	*					
G. lucidum			*		*		*
G. mastoporum			*				
Gloeoporus chroceopallens					*		
G. dichrous					*		*
G. thelephoroides	*	*			*	*	*
Hapalopilus albocitrinus					*		
Hexagonia hydnoides	*				*		
H. tenuis	*	*	*	*	*		*
Junghuhnia nitida		*	*				
Laetiporus sulphureus							*
Lenzites acuta							*
L. elegans							*
Loweporus inflexibilis					*	*	
L. roseoalbus		*			*		
Megasporoporia cavernulosa	*	*					
M. setulosa	*	*					
Microporellus obovatus				*	*		
Microporus affinis			*	*	*		
M. microloma			*			*	*
M. vernicipes	*	*	*	*	*		*
M. xanthopus	*	*	*	*	*	*	*

Continued

Table 6.2. (*Continued*).

Species	Bali monsoon	Bali lowland	Bali cloud	Lombok lowland	Lombok cloud	Sumbawa monsoon	Sumbawa cloud
Nigrofomes melanoporus	*	*	*	*	*		*
Nigroporus vinosus				*	*		
Oligoporus subcaesius		*	*	*	*		
Oxyporus cervinogilvus						*	*
O. populinus						*	
Perenniporia contraria							*
P. medulla-panis			*				
P. ochroleuca				*			
P. tephropora		*	*				*
Phellinus allardii						*	*
P. callimorphus							*
P. discipes						*	
P. fastuosus						*	*
P. gilvus	*						
P. glaucescens						*	*
P. lamaensis			*		*		*
P. noxius							*
P. pachyphloeus							*
Phylloporia chrysita	*	*		*			
P. spathulata				*			
Polyporus arcularius					*	*	*
P. dictyopus		*	*		*		
P. grammocephalus		*	*	*			*
P. tenuiculus	*	*	*	*	*		
P. udus							*
Protomerulius caryae					*		

Pseudofavolus cucullatus	*						
P. miquelii	*	*					
Pycnoporus sanguineus	*	*					*
Rigidoporus dextrinoideus					*		
R. lineatus			*		*		
R. microporus			*	*			
R. vinctus		*	*		*		*
Schizophyllum commune					*		
Schizopora flavipora		*	*	*			*
S. paradoxa		*	*			*	
Skeletocutis diluta			*	*	*	*	*
S. lenis							*
S. nivea					*	*	*
Theleporus calcicolor					*		*
Tinctoporellus epimiltinus		*	*	*	*		*
Trametes cingulata		*	*			*	
T. lactinea	*						
T. menziezii		*					*
T. mimetes		*					
T. socotrana					*		
T. versicolor			*		*		
Trichaptum durum		*	*	*	*	*	*
Wrightoporia tropicalis							*

Table 6.3. World distribution of polypores collected in the Lesser Sunda Islands. The highest and lowest scores are shown in bold.

	Number	Percentage
Cosmopolitan	18	17.5
Pantropical	**52**	**51**
Palaeotropical	25	24.5
Australasian	2	2
Asian	3	3
Australian	**0**	**0**
Holarctic	2	2
Total	102	

The total number of species collected in each locality and the unique component, that is, the subset of species found in each locality and nowhere else, are shown in Table 6.4. Total numbers ranged from 45 species sampled in Sumbawa cloud forest to 23 species in Bali monsoon forest. Only three species were common to all localities: *Hexagonia hydnoides* (Schwartz) M. Fidalgo, *Microporus vernicipes* (Berk.) Kuntze and *Microporus xanthopus* (Fr.) Kuntze.

Similarity Indices

The similarity indices and numbers of species shared between the sampled localities are shown in Table 6.5. The highest similarity index (0.59) and highest number of shared species (24) were found between the cloud forests of Bali and Lombok. The lowest similarity index (0.19) was found between the lowland forest of Lombok and the monsoon forest of Sumbawa.

Similarity indices for the three Lesser Sunda Islands and Java + Sumatra, a combination representative of the Asian plate, have been compared, and also with Australia + PNG, a combination representative of the Australian plate (Table 6.6). In Table 6.6, the three Lesser Sunda Islands are first shown as a single local-

Table 6.4. Total species numbers and unique components (i.e. species found in one locality and nowhere else) for each locality studied in the Lesser Sunda Islands. The highest and lowest species numbers are shown in bold.

	Total	Unique component
Bali monsoon	**23**	4
Bali lowland	39	6
Bali cloud	36	**3**
Lombok lowland	25	**3**
Lombok cloud	44	9
Sumbawa monsoon	28	8
Sumbawa cloud	**45**	**10**

Table 6.5. Similarity indices between the different collecting localities. The highest and lowest scores are shown in bold. The number of shared species is indicated in parentheses.

	Bali lowland	Bali cloud	Lombok lowland	Lombok cloud	Sumbawa monsoon	Sumbawa cloud
Bali monsoon	0.55(17)	0.27(8)	0.33(8)	0.24(8)	0.24(6)	0.35(12)
Bali lowland		0.58(15)	0.47(15)	0.41(17)	0.23(8)	0.45(19)
Bali cloud			0.52(16)	**0.59(24)**	0.22(7)	0.32(16)
Lombok lowland				0.49(17)	**0.19(5)**	0.28(10)
Lombok cloud					0.22(8)	0.45(20)
Sumbawa monsoon						0.41(15)

ity, then they have been divided according to their placement on the Asian (Bali) or the Australian (Lombok + Sumbawa) tectonic plates. As a single locality, the Lesser Sunda Islands have higher similarity indices when compared with Java + Sumatra (0.53) than when compared with Australia + PNG (0.40). However, the highest similarity index was obtained between Bali and Lombok + Sumbawa (0.55, Table 6.6).

Discussion

Species composition in Bali, Lombok and Sumbawa

The results of these surveys should be regarded as preliminary, as much collecting remains to be done in the area. This is particularly applicable to the non-volcanic islands of the southern archipelago of the Lesser Sunda Islands, including Sumba, Timor and Babar.

No sharp demarcation line for polypore distribution can be drawn between Bali, on the Asian plate, and Lombok + Sumbawa, on the Australian plate, which accords with previous work on angiosperms (van Steenis, 1979) and mosses (Touw, 1992). As a general pattern, species numbers were highest in

Table 6.6. Similarity indices in the Pacific Australasian region. The Lesser Sunda Islands are treated as a single locality (LSI), or as two localities on either the Asian or the Australian plate (i.e. Bali or Lombok or Sumbawa). PNG, Papua New Guinea. The highest value is shown in bold, and the number of shared species in parentheses.

	LSI	Bali	Lombok + Sumbawa	Australia + PNG
Java+Sumatra	0.53(69)	0.43(47)	0.46(56)	0.51(116)
Bali			**0.55(40)**	0.30(54)
Lombok + Sumbawa				0.33(64)
LSI				0.40(81)

the cloud forests and lowest in the monsoon forests. The low species numbers in the monsoon forests were probably largely attributable to collecting having been at the end of the dry season both in 1994 and in 1996.

Species unique to each locality were highlighted in an attempt to identify any endemic elements in the area. However, these unique components were always of either Asian or palaeotropical character and therefore the majority are likely to be found in other localities in the area when more collecting is undertaken, especially during the rainy season.

Similarity indices

The similarity indices obtained from Bali, Lombok and Sumbawa do not indicate any clear difference between the polypore mycota on either side of Wallace's line. The highest similarity index was obtained between Bali and Lombok cloud forests, on opposite sides of Wallace's line (Table 6.5). The polypore mycota of Bali, Lombok and Sumbawa was more similar to that of Java + Sumatra (Asian plate) than to that of Australia + PNG (Australian plate), regardless of whether the three Lesser Sunda Islands were considered as a single locality or as two localities separated by Wallace's line (Table 6.6).

Quanten (1993) found that there were twice as many Asian polypore species in PNG than Australian ones, even if PNG lies to the east of Wallace's line. She attributed this fact to the distribution of the higher plant vegetation types: 'the Indo-Malayan region is dominated by rainforest, while only a small area of Australia is covered by this vegetation type'.

In general, the highest similarity indices were obtained when comparing either lowland rainforest with cloud forest localities, or two cloud forest localities (Table 6.5). This indicates that the forest type is a more important factor for species composition than the islands' geographic position relative to Wallace's line.

Van Steenis (1984) established a floristic zonation in Malesia according to altitude. He defined a tropical zone (0–1000 m), a submontane zone (1000–1500 m), and a montane zone (1500–2400 m). Our study shows that polypore distribution does not follow the same zonation as angiosperms, since localities showing highest similarity indices ranged from 500 to 1700 m altitude (Tables 6.1 and 6.5).

Sumbawa monsoon forest had the lowest similarity indices when compared with other localities, a preliminary result which, due to limited collecting, should be treated with caution.

Species distribution in Pacific Australasia

The species lists compiled for the Pacific Australasian regions of Java, Sumatra, Australia and PNG showed that both Asia and Australia can be characterized by groups of species that do not cross Wallace's line. Asian examples include *Daedalea incana* (Lév.) Ryv., *Haddowia aetii* Steyaert, *Humphreya endertii* Steyaert, *Hymenogramme javanensis* Mont. & Berk. and *Polyporus pervadens* Corner, whereas examples from Australia include *Australoporus tasmanicus* (Berk.) P.K. Buchanan & Ryv., *Dichomitus epitephrus* (Berk.) Ryv., *Dichomitus leucoplacus* (Berk.) Ryv., *Laccocephalus sclerotinus* (Rodway) Núñez & Ryvarden and *Piptoporus australiensis* (Wakef.) G. Cunn. Other species, for example *Diacanthodes novoguienensis* (Henn.) O. Fidalgo and *Macrohyporia dictyopora* (Cooke) I. Johans. & Ryvarden, appear to have a Gondwanaland origin, as they occur in both Australia and South America. Hjortstam and Ryvarden (1985) and Rajchenberg (1989) have discussed the Gondwanaland element in South America and Australia.

In contrast, the Pacific Australasian lists also show that several species do cross Wallace's line. However, due to the lack of fossil polypores, the time and direction of migration have to be inferred from other indirect methods. Croizat (1958) established the 'generalized track' theory to explain species distribution in an area. According to this theory, the most probable distribution patterns are those shared by several organism groups and concord with the known geological history of an area. Several distribution patterns have been suggested for Pacific Australasia (Touw, 1992; Michaux, 1994) and several polypore species can be tentatively assigned to them as follows:

1. *Temperate Australasian species.* Parts of the present continental Asia, e.g. Myanmar, western Thailand, Malay Peninsula and Sumatra, were in Gondwanaland and drifted towards Laurasia during the Jurassic (Audley-Charles, 1987). Burbidge (1960) stated that several Australian temperate plants apparently migrated with the land. A similar event may have happened with several temperate Australian polypores, for example *Coltriciella tasmanica* (Cleland & Rodway) D.A. Reid, *Heterobasidion insulare* (Murrill) Ryv., *Inonotus setulosocroceus* (Cleland & Rodway) P.K. Buchanan & Ryv. and *Laccocephalum hartmanni* (Cooke) Núñez & Ryv., which are known from the Asian temperate mycota in Japan, China and far eastern Russia. However, none of these species was found in the Lesser Sunda Islands. It is important to note that none of these Australian polypore species is known from southern South America, therefore they may have evolved after the Australian–Antarctic plate separated from the rest of Gondwanaland in the middle Eocene, some 45 million years ago (Schuster, 1972).

2. *Disjunct species within continental Asia.* During the Pliocene glaciation maxima, species at higher latitudes in continental Asia, mostly from warm-temperate areas, migrated southwards. The sea level had decreased about 200 m and the montane vegetation limit had also lowered (Audley-Charles, 1987). As more humid conditions returned to the islands after the glaciations, the rainforests gained terrain in the lowlands. Several of the warm-temperate species were isolated in the mountains and are now scattered over Wallace's line in the cloud and wet forests of Lombok and Sumbawa; for example *Coriolopsis glabrorigens*

(C.G.Lloyd) T. Hatt., *Echinoporia hydnophora* (Berk. & Broome) Ryv., *Gloeoporus croceopallens* Bres. and *Hapalopilus albocitrinus* (Petch) Ryv.. As is the case for mosses showing the same distribution pattern (Touw, 1992), the above species are adapted to drought, and appear to require a dry period to complete their life cycle.
3. *Species crossing Wallace's line in both directions as the Australian continent approached Asia from the mid-Miocene* (Raven and Axelrod, 1972). Examples among the polypores include *Ceriporia ferruginicincta*, *Cerrena meyenii* (Klotzsch) L. Hansen, *Echinochaete ruficeps* (Berk. & Broome) Ryv., *Elmerina cladophora*, *Mollicarpus cognatus* (Berk.) Ginns, *Oxyporus cervinogilvus* and *Phellinus glaucescens* (Petch) Ryv.. Intensive and extensive collecting in the Lesser Sunda Islands, especially in those of the southern archipelago, is critical to determine the direction of migration. Most of the species mentioned above are of tropical character. In other organism groups, migration of tropical species appears to be mainly from the Asian rainforests to those of East Australia (Michaux, 1994). Of the species mentioned above, only *M. cognatus* has not been found in the Lesser Sunda Islands.

Conclusions

The polypore mycota in several Lesser Sunda Islands, southern Indonesia, on both sides of Wallace's line, Bali on the Asian plate, and Lombok and Sumbawa on the Australian plate, have been compared. The results are extrapolated to include known distribution patterns of polypores in temperate and tropical East Asia and Australia.

Most of the collected species in Bali, Lombok and Sumbawa (77%) were either pantropical or palaeotropical. The polypore mycota did not differ significantly on either side of Wallace's line. The type of forest ecosystem seems to be a more important factor in distribution patterns of polypores than the geographical situation with respect to Wallace's line.

On a wider geographical scale, several species were tentatively assigned to the following distribution tracks in Australasia:

1. Australian temperate species that were rifted to East Asia during the Cretaceous: *C. tasmanica*, *H. insulare*, *I. setulosocroceus*, *L. hartmanni*, these species were not found in the Lesser Sunda Islands.
2. Vicariant species with continental Asia after the Pliocene glaciation maxima, present today as a discrete group in the cloud forests of Lombok and Sumbawa: *C. glabrorigens*, *E. hydnophora*, *G. croceopallens*, *H. albocitrinus*.
3. Species transgressing Wallace's line in both directions as the Australian continent approached Asia from the mid-Miocene: *C. ferruginicincta*, *C. meyenii*, *E. ruficeps*, *M. cognatus*, *O. cervinogilvus*, *P. glaucescens*.

C. ferruginicincta and *S. diluta* are new records for Asia.

Acknowledgements

Thanks are due to the British Mycological Society, especially Professor Roy Watling, for inviting us to present this work at the BMS Millennium Meeting in Liverpool. We warmly thank the former Director of Bali Botanical Gardens, Mr I.B.K. Arsana, for his hospitality during our stay in Bali, and Mr Tjuk Sukarto, Mr Cecep Suryana, Mr Cecep Rayadi, Mr I. Gusti Made Radhitha, and Mr Edy for their efficient help with logistics during the expeditions.

References

Audley-Charles, M.G. (1987) Dispersal of Gondwanaland: relevance to evolution of the Angiosperms. In: Whitmore, T.C. (ed.) *Biogeographical Evolution of the Malay Archipelago*. Oxford Monographs on Biogeography 4, Clarendon Press, Oxford, pp. 5–25.

Burbidge, N.T. (1960) The phytogeography of the Australian region. *Australian Journal of Botany* 8, 75–211.

Croizat, L. (1958) *Panbiogeography*, Vols 1, 2a, 2b. Croizat, Caracas, Venezuela.

Cunningham, G.H. (1965) Polyporaceae of New Zealand. *New Zealand Department of Sciences and Industrial Research*, Wellington,164, 1–304.

Hisheh, S., Westerman, M. and Lincoln, H.S. (1998) Biogeography of the Indonesian archipelago: mitochondrial DNA variation in the fruit bat, *Eonycteris spelaea*. *Biological Journal of the Linnean Society* 65, 329–345.

Hjortstam, K. and Ryvarden, L. (1985) New and noteworthy basidiomycetes (Aphyllophorales) from Tierra del Fuego, Argentine. *Mycotaxon* 22, 159–167.

Imazeki, R. (1952) A contribution to the fungous flora of Dutch New Guinea. *Bulletin of the Government Forest Experiment Station* 57, 87–128.

Michaux, B. (1994) Land movements and animal distributions in East Wallacea (eastern Indonesia, PNG and Melanesia). *Palaeogeography, Palaeoclimatology, Palaeoecology* 112, 323–343.

Musser, G.G. (1981) The giant rat of Flores and its relatives East of Borneo and Bali. *Bulletin of the American Museum of Natural History* 169, 71–175.

Pirozynski, K.A. (1983) Pacific mycogeography: an appraisal. *Australian Journal of Botany* Supplement Series 10, 137–159.

Quanten, E. (1993) Study of the Polyporaceae s.l. of Papua New Guinea: preliminary polypore flora. PhD thesis, The University of Limburg, Belgium.

Rajchenberg, M. (1989) Polyporaceae (Aphyllophorales, Basidiomycetes) from Southern South America: a mycogeographical view. *Sydowia* 41, 277–291.

Raven, P.H. and Axelrod, D.I. (1972) Plate tectonics and Australasian paleobiogeography. *Science* 176, 1379–1385.

Ryvarden, L. (1981) Type studies in the Polyporaceae 12. Species described by F.W. Junghuhn. *Persoonia* 11, 369–372.

Schuster, R.M. (1972) Continental movements, 'Wallace's line' and Indomalayan-Australasian dispersal of land plants: some eclectic concepts. *The Botanical Review* 38, 3–86.

Simpson, G.G. (1977) Too many lines: the limits of the oriental and Australian zoogeographic regions. *Proceedings of the American Philosophical Society* 121, 107–120.
Sims, K., Watling, R., de la Cruz, R. and Jeffries, P. (1997) Ectomycorrhizal fungi of the Philippines: a preliminary survey and notes on the geographic biodiversity of the Sclerodermatales. *Biodiversity and Conservation* 6, 45–58.
Steenis, C.G.G.J. van (1979) Plant-geography of East Malesia. *Botanical Journal of the Linnean Society* 79, 97–178.
Steenis, C.G.G.J. van (1984) Floristic altitudinal zones in Malesia. *Botanical Journal of the Linnean Society* 89, 289–292.
Steyaert, R.L. (1972) Species of *Ganoderma* and related genera mainly of the Bogor and Leiden herbaria. *Persoonia* 7, 55–118.
Suhirman and Núñez, M. (1995) Indonesian Aphyllophorales I. *Buletin Kebun Raya Indonesia (Indonesian Botanic Garden Bulletin)* 8, 69–78.
Suhirman and Núñez, M. (1996) Indonesian Aphyllophorales II. Species of Polyporaceae in the Herbarium Bogoriensis. *Buletin Kebun Raya Indonesia (Indonesian Botanic Garden Bulletin)* 8, 107–109.
Suhirman and Núñez, M. (1998) Indonesian Aphyllophorales 3. Poroid and stereoid species from Kerinci-Seblat National Park, Western Sumatra. *Mycotaxon* 68, 273–292.
Touw, A. (1992) Biogeographical notes on the Musci of South Malesia, and of the Lesser Sunda Islands in particular. *Bryobrothera* 1, 143–155.
Veevers, J.J. (1991) Phanerozoic Australia in the changing configuration of Proto-Pangea through Gondwanaland and Pangea to the present dispersed continents. *Australian Systematic Botany* 4, 1–11.
Wallace, A.R. (1860) On the zoological geography of the Malay Archipelago. *Zoological Journal of the Linnean Society London* 4, 172–184.

The Biology, Ecology and Pathology of *Phellinus noxius* in Taiwan 7

T.T. Chang

Division of Forest Protection, Taiwan Forestry Research Institute, 53 Nan-hai Road, Taipei 100, Taiwan

Introduction

Phellinus noxius (Corner) Cunn. is widely distributed in tropical regions in South-east Asia, Oceania and Africa (Larsen and Cobb-Poulle, 1990), where it causes a brown root disease that is responsible for the decline of numerous orchard and forest tree species (Singh *et al.*, 1980; Turner, 1981; Bolland, 1984; Hodges and Tenorio, 1984; Neil, 1986; Nandris *et al.*, 1987). In Taiwan in the 1950s, tea, coffee and 15 other plant species were documented as hosts of *P. noxius* based on the presence of a dark brown mycelial mat on the surface of diseased roots (Tsai, 1991). However, the causal fungus was not isolated and further studies of brown root disease were discontinued until the late 1980s because *P. noxius* basidiomes were rarely observed on host plants in the field. More recently, it has been isolated from woody hosts and its basidiomes have been produced on sawdust medium (Ann and Ko, 1992; Chang, 1992), thus confirming an association between the fungus and brown root disease on woody plants. Consequently, there has been a renaissance of studies on *P. noxius* and brown root disease during the last decade.

Phellinus noxius in Laboratory Culture

Well-developed basidiomes were rarely observed on diseased trees in the field, so the characteristics of pure cultures and laboratory-produced basidiomes were used for identification of *P. noxius*.

The fungus was easily isolated from diseased roots and stems by using a selective medium consisting of malt extract (20 g), agar (20 g), benomyl (10 mg), dicloran (10 mg), ampicillin (100 mg) and gallic acid (500 mg) l^{-1} (Chang, 1995a). Cultures were initially whitish on malt extract agar and potato dextrose agar (PDA) becoming brown with irregular dark brown zone lines or patches permeating the culture. Arthroconidia and trichocysts were always present in the culture, while clamp connections were absent on generative hyphae. The arthroconidia were hyaline and rod-shaped and trichocysts were dark brown. The only specialized elements were yellow to rust-brown skeletal hyphae. Within the genus *Phellinus*, the combination of arthroconidia and trichocysts is only observed in cultures of *P. noxius* and is therefore very useful in the identification process. Examination of the fungus from diseased plants revealed that trichocysts were commonly present on the diseased tissues, but arthroconidia were rare or not observed.

Cultures isolated from 15 hosts showed similar growth responses to different temperature regimes, with an optimum temperature near 30°C, a maximum temperature above 36°C, and a minimum temperature of 10–12°C. However, the growth rates of different isolates at the optimum growth temperature varied greatly. The linear growth rates ranged from 3.4 to 0.8 cm day^{-1} (Chang, 1992; Ann *et al.*, 1999a). Mycelia of the tested isolates grew in PD broth with original pH values ranging from 3.5 to 7.0, but no growth was observed above pH 7.5 (Ann *et al.*, 1999a).

Tests for extracellular oxidases were positive for laccase and peroxidase but not for tyrosinase, indicating that *P. noxius* belongs to the group of white rot fungi (Chang, 1992). Chang *et al.* (1994) found that *P. noxius* did not secrete nitrate reductase and thus was unable to use $NaNO_3$, KNO_3 and $Ca(NO_3)_2$ as nitrogen sources. However, this species used ammonium salts such as NH_4NO_3, NH_4Cl, $(NH_4)_2HPO_4$ and $(NH_4)_2SO_4$, and organic nitrogen such as asparagine and urea.

Basidiome formation and morphology

In the field, basidiomes of *P. noxius* have been observed on only nine out of 101 host plant species, namely *Annona squamosa* L., *Casuarina equisetifolia* L., *Cinnamomum camphora* (L.) Nees et Eberm., *Delonix regia* (Boj.) Raf., *Euphoria longan* Lam., *Ficus elastica* Roxb., *Ficus microcarpa* L., *Litchi chinensis* Sonn. and *Melicope merrilli* (Kaneh. et Sasak. ex Kaneh.) Liu et Liao (Chang and Yang, 1998; Ann *et al.*, 1999b). However, isolates from all hosts produced basidiomes on a hardwood sawdust medium consisting of sawdust (40 kg), rice bran (10 kg), sucrose (50 g), NH_4NO_3 (10 g), citric acid (5 g) and about 15% (w/w) water in plastic bags (Chang and Yang, 1998). Basidiomes of *P. noxius* produced on sawdust medium were usually resupinate, whereas those observed in the field were resupinate to effused–reflexed. Although the macromorphology of *P. nox-*

ius varied between cultured and naturally occurring basidiomes, the micromorphology was similar. The hyphal system was dimitic comprising generative and skeletal hyphae. Generative hyphae were 2–4 μm in diameter, hyaline to yellow and lacked clamp connections. Skeletal hyphae were yellow-brown to bay with a diameter of 3–6 μm. Contextual and tramal setal hyphae were dark ferruginous and 370–602 × < 13 μm. Projection of the latter into the hymenium was difficult to observe. Basidiospores were smooth, hyaline, broadly ellipsoid to subglobose, and measured 3–4 × 4–7 μm.

Phellinus noxius in the Field

Disease pattern and dissemination

In recent years, the brown root disease caused by *P. noxius* has become one of the most serious problems of fruit and forest trees at lower elevations in central and southern Taiwan (Ann and Ko, 1992; Chang, 1992, 1995b). Several slowly expanding, circular disease patches extending from infection centres were observed in the field indicating spread from diseased to healthy trees by root contact. New disease patches were not commonly observed in the field indicating that airborne basidiospores and arthroconidia are not efficient in the establishment of new infections at their natural levels of production. Nevertheless, airborne basidiospores can initiate infections on freshly cut stumps or infest logging debris with subsequent spread to live trees by root contact (Anonymous, 1974a,b; Singh *et al.*, 1980; Turner, 1981; Hodges and Tenorio, 1984).

The survival of *Phellinus noxius*

The mycelium, including that within colonized wood, arthroconidia and basidiospores, was measured in soils with different matrix potentials (Chang, 1996). Survival of arthroconidia declined more slowly in soils with matrix potentials of –0.50 and –0.42 MPa than in soils with potentials of –0.15 and –0.025 MPa. However, arthroconidia were rarely recovered from the treatments after 3 months (Fig. 7.1). Basidiospores were not recovered after 3.5, 4, 4.5 and 5 months in the –0.025, –0.15, –0.42 and –0.50 MPa soil matrix potential treatments, respectively (Fig. 7.2). Mycelia which had colonized cellophane membranes buried in soil with potentials of –0.025 MPa were not recovered after 6 weeks, whereas mycelia on cellophane in soils of –0.50, –0.42 and –0.15 MPa were not recovered after 12 weeks (Fig. 7.3). *P. noxius* was not recovered from inoculated pieces of wood subjected to 1 month of flooding. However, in treatments with lower soil moisture, *P. noxius* survival ranged from 80 to > 90% over 2 years (Fig. 7.4). Colonies of *P. noxius* were not recovered from the rhizosphere of soils around the colonized roots of three host species: *Calophyllum inophyllum* L., *C. equisetifolia*, and *C. camphora*. However, *P. noxius* was recovered from naturally colonized roots of these hosts 1–10 years after they were killed. These results indicate that woody debris in soils harbouring *P. noxius* plays an important role in the long-term survival of the fungus.

Although the longest recorded survival time of *P. noxius* in dead roots was 10 years after the host plant was killed, it appeared that the fungus could survive longer based on approximately 50% survival over 10 years (Chang, 1996). The survival pattern of *P. noxius* on different hosts over a long period of time might show differences, but only 10-year-old dead trees were available in the study. It was difficult to find roots that had been dead for longer periods because

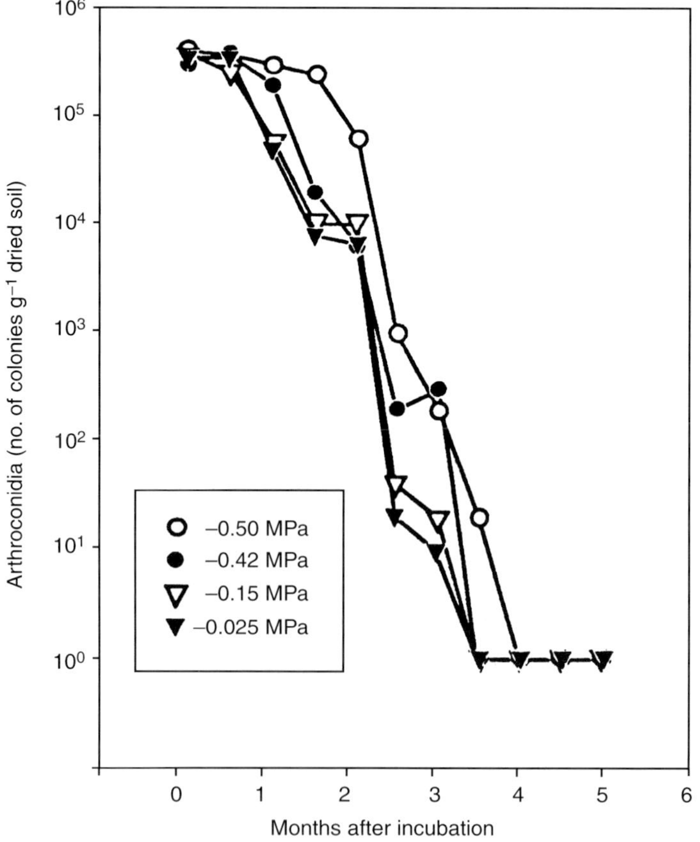

Fig. 7.1. Survival of *P. noxius* arthroconidia mixed in soils with varying soil moisture treatments at different incubation times. Each point is the mean of six replicates.

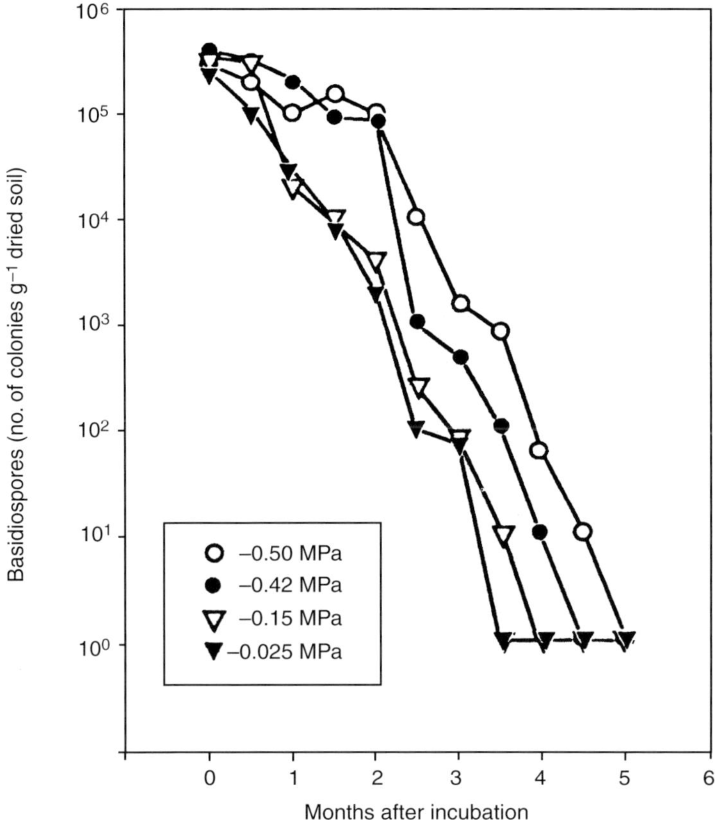

Fig. 7.2. Survival of *P. noxius* basidiospores mixed in soils with varying soil moisture treatments at different incubation times.

the disease has not been studied carefully until recently. After the death of its host, *Phellinus weirii*, which causes laminated root rot of *Pseudotsuga menziesii* (Mirb.) Franco (Douglas fir) in the western United States and Canada, continues to live saprobically in the lower portion of the trunk and the root system for 50 years or more (Thies, 1984). The possibility that chlamydospores may play a role in the long-term survival of *P. noxius* was investigated. Previous work did not reveal chlamydospores in pure cultures of *P. noxius* (Anonymous, 1974b) and microscopic examination of colonized soils similarly did not reveal their presence (Chang, 1996). Based on these results, it was concluded that *P. noxius* does not form chlamydospores as a long-term survival strategy. However, *P. noxius* was able to live in woody debris in soil until the wood had rotted away.

Viability of basidiospores, arthroconidia and mycelia declined rapidly in soil compared with that of the fungus in colonized woody debris. Indeed, only the latter harboured the fungus in a viable state after 5 months. It was possible that

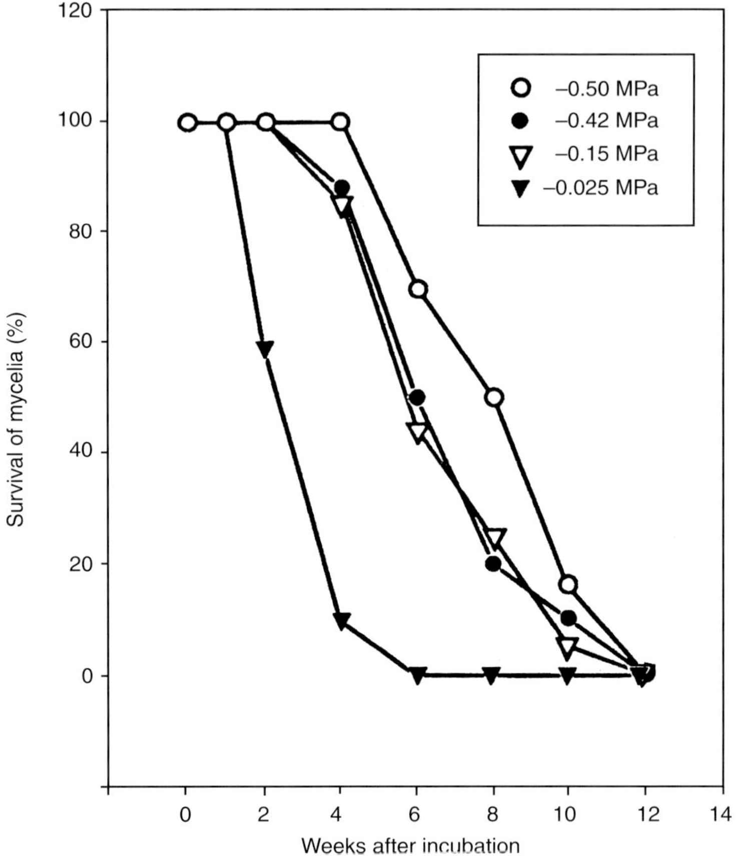

Fig. 7.3. Survival of *P. noxius* mycelia on cellophane buried in soils varying with soil moisture treatments at different incubation times.

populations of basidiospores and arthroconidia of low viability existed in soils after 5 months although they were not recoverable by the methods used in the study. Basidiospores are not generally regarded in a long-term survival context, but they play an important role in the long distance dissemination of *P. noxius*. Infection of trees, presumably by airborne basidiospores, has been reported for cacao and rubber (Anonymous, 1974b).

Submerging wood in water has been used to prevent postharvest colonization by wood-decaying fungi. In this study, when *P. noxius*-colonized wood sections were submerged in water, the fungus died within 1 month. This might have involved the activities of anaerobic microorganisms, but further study is required to determine the exact mechanism. Flooding colonized fields for more than 1 month might eliminate the inoculum source in colonized wood and roots. However, field studies are required to test this hypothesis and may lead to

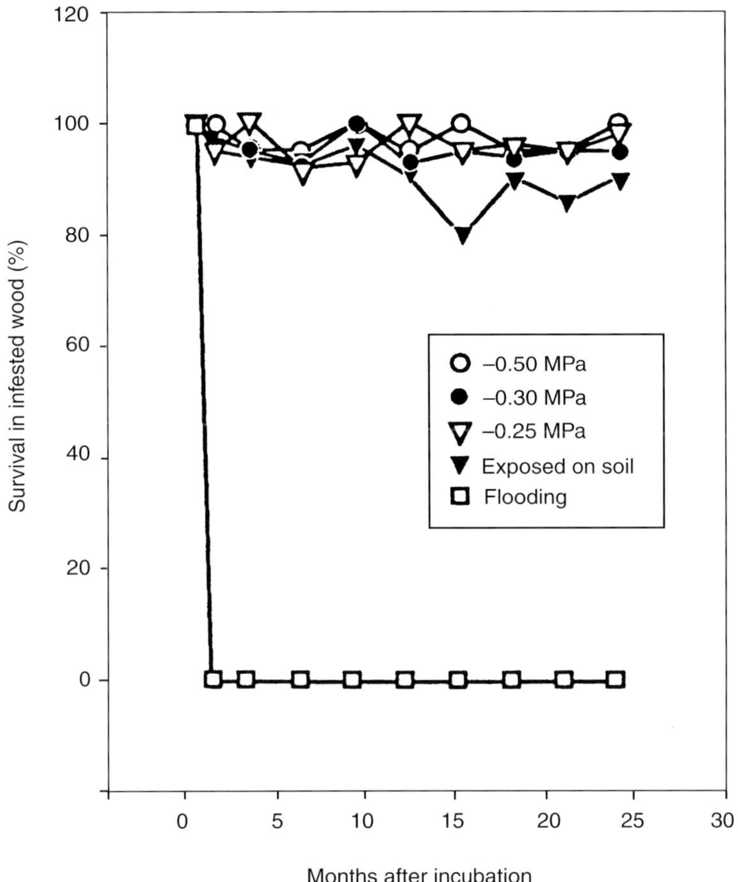

Fig. 7.4. Survival of *P. noxius* in artificially infected wood sections buried in soils with varying moisture treatments at different incubation times. Each point is the mean of six replicates. The least significant difference test among all treatments at different incubation times when the flooding treatment was excluded showed that the means at only three incubation times were significantly different ($P = 0.01$) from other treatments.

more effective field control practices. According to a disease survey, *P. noxius* occurred more frequently in fields with sandy soils, probably due to the better drainage and reduced likelihood of flooding even during high rainfall. Clay soils drain poorly and therefore these soils are more likely to be submerged, a situation that is detrimental to the survival of *P. noxius*.

Distribution and environment

P. noxius has been collected at 260 different sites in Taiwan, and two different sites on Chinmen Island, which is much closer to Fujian Province, China, than to Taiwan. Collections were mostly from host plants growing in the low elevation plains and hills of western, eastern and southern Taiwan, while it was found on only a few hosts in northern Taiwan and was absent from the high mountainous areas. Of the 262 localities, 240 were from the plains with elevations < 500 m, while 22 localities were from low mountainous areas of 500–1000 m elevation (Chang and Yang, 1998; Ann *et al.*, 1999b).

P. noxius was observed mostly in central and southern Taiwan, in low elevation areas, indicating that the disease is primarily tropical and subtropical in distribution. This agrees with the results of previous studies (Hodges and Tenorio, 1984; Neil, 1986; Nandris *et al.*, 1987; Ann and Ko, 1992; Chang, 1992, 1995a). Only a few cases of the fungus disease were present in northern Taiwan. This could be as a result of rainfall distribution rather than temperature because the lowest and highest temperatures in central and northern Taiwan do not differ. The fungus was also found on Chinmen Island where the temperature is lower than in northern Taiwan. However, the winter monsoon brings long periods of rain and high humidity to northern Taiwan, while the other regions are usually dry.

Soil samples (153) collected from the rhizospheres of the diseased plants were used for determination of soil pH and texture. Although *P. noxius* did not favour a particular soil pH, it was not found in extreme pH conditions. It was found in soils with pH 4–9, and most frequently in soils with pH 5–8. Diseased plants were not found in soils < pH 4 or > pH 9. *P. noxius* occurred mostly in soils classified as sand, loamy sand and sandy loam, although it was occasionally found in other types of soils.

This study also demonstrated that soil pH has no significant effect on *P. noxius* and the effect of pH may be expressed through the host rather than directly on the fungus. This phenomenon has also been shown by *Armillaria mellea* (Vahl: Fr.) Kummer and *P. weirii* (Murr.) Gilbn., two important root rot fungi in temperate forests (Angwin, 1985; Shaw and Kile, 1991). Soil pH appears to have no significant effect on *P. noxius* in Taiwan, but the fungus is more common in sandy soils. The latter may be a feature of drainage: sandy soils are generally well drained and other evidence suggests that dry soil conditions are favourable to the survival of *P. noxius* (Chang, 1996). Moreover, the percentage of trees infected by *Armillaria ostoyae* (Romagnesi) Herink in Ontario, Canada, was higher at a dry site than at a wet site (Whitney, 1988). Additionally, the disease caused by *Heterobasidion annosum* (Fr.) Bref. was found to occur more frequently in drier sites with sandy soil (Rishbeth, 1950).

Symptoms and host ranges

The disease symptoms caused by *P. noxius* progressed from discoloration of leaves to gradual defoliation, slow growth and eventual death of the infected trees. A dark brown mycelial mat formed on the surface of the roots and up the base of the stem but usually no further than 1 m above the ground. Masses of sand and stone usually adhered to the mycelium on the root and stem surface. Dead colonized bark was brittle and easily removed, and its inner surface was covered with white to brownish mycelial mats with a network of mycelial cords. The colonized woody tissues eventually decomposed and showed typical white rot symptoms. The diseased trees also died when the infection reached the root crown.

P. noxius was identified from 101 woody plants, comprising 91 hardwoods and 10 conifers, and four annual herbaceous plants. The four annuals were found in a Taichung harbour windbreak of *C. equisetifolia* which was heavily infected with *P. noxius*. In this case, the fungus caused brown root rot but did not severely wilt the foliage (Chang and Yang, 1998; Ann *et al.*, 1999b).

In order to determine the variation in host susceptibility and virulence of isolates from various hosts, 12 tree species were inoculated in all possible combinations with an isolate from each of the hosts (Chang, 1995b). All isolates of *P. noxius* used were able to cause disease in all 12 tree species. However, disease incidence was quite variable, ranging from 20 to 90% of the trees. Survival of *Acacia confusa* Merr. and *Salix babylonica* L. was significantly better than that of other host plants using logistic regression analysis. Indeed the incidence of disease was significantly lower on these two tree species, indicating that they may have some level of resistance or tolerance to the disease. In contrast, there were no significant differences among fungal isolates. Cross pathogenicity tests demonstrated that *P. noxius* had a wide host range without showing host specificity for any of the species tested. However, different degrees of resistance to the fungus might be present in different tree species. As has been observed in the field, some trees such as *Melaleuca leucadendra* (L.) L. and *Alstonia scholaris* R. Br. were found to be free of disease when they were growing in close proximity to diseased trees of other species. Hence, further study is required to determine whether resistant tree species exist.

Fungicidal Effect of Urea on *P. noxius* in Alkaline Soil

Removal of colonized stumps and roots from the soil to reduce the source of inoculum has been widely used to control diseases caused by root rot fungi (Kuhlman *et al.*, 1976; Wallis, 1976; Thies *et al.*, 1994). This practice is probably more effective for *P. noxius* than for other root rot fungi because *P. noxius* rarely produces basidiospores in the natural environment (Chang, 1995b). However, small roots are easily broken off and left in the soil during uprooting. These infected pieces usually act as primary inocula and spread the disease. Thus, control practices that reduce or eradicate *P. noxius* from colonized root residue are essential for effective control of the disease. Urea proved to be effective at killing *P. noxius* in colonized wood fragments (Chang and Chang, 1999). *P. noxius* was not recovered from pieces of inoculated wood placed in or on soil amended with urea, $(NH_4)_2CO_3$ or aqueous ammonia. High concentrations of volatile NH_3 were detected in these treatments, indicating that NH_3 generated from these chemicals was fungicidal to *P. noxius* (Table 7.1). A high concentration of volatile NH_3 was detected from non-autoclaved soil amended with urea but not from amended, autoclaved soil, indicating that soil microorganisms were involved in NH_3 generation. To kill *P. noxius* completely required 3000 p.p.m. urea in soil. Volatile NH_3 was generated from urea in alkaline soil but not in acidic soil (Table 7.2). Similar quantities of volatile NH_3 were generated at temperatures ranging from 12 to 32°C. A large quantity of NH_3 was generated at low soil matrix potentials (–0.75 to –0.15 MPa), while high potentials (> –0.025 MPa) and flooded soil hindered production of volatile NH_3 (Table 7.3). Ammonia was more effective at killing *P. noxius* growing in smaller pieces of wood than in larger pieces. The fungus was not recovered from wood pieces less than 3 cm in diameter. Volatile NH_3 was also lethal to six other root rotting fungi: *Ganoderma australe* (Fr.) Pat., *Ganoderma lucidum* (Curt.:Fr.) Karst., *Ganoderma tropicum* (Jungh) Bres., *Rigidoporus vinctus* (Berk.) Ryv., *Heterobasidion annosum* and *Rosellinia necatrix* Prill.

This wide spectrum of fungicidal activity is similar to that of soil fumigants and could be enhanced by covering the ground with plastic sheeting. Urea is an inexpensive fertilizer and has relatively low toxicity to plants, especially to perennial woody plants. Thus, urea holds promise as an agent for controlling *P. noxius* and other fungi that cause disease of woody plants.

Table 7.1. Effects of urea and other chemicals on volatile NH_3 generation and *P. noxius* survival in inoculated wood sections buried in soil.

		Survival of *Phellinus noxius* (%)		Concentration of
	p.p.m.	In soil	On soil	NH_4^+ (mg l^{-1})
Urea				
Autoclaved soil	0	100[a2]	100[a]	5.3[b]
	1000	100[a]	100[a]	6.4[b]
Non-autoclaved soil	0	100[a]	100[a]	8.5[b]
	1000	100[a]	100[a]	87.4[b]
	3000	0[b]	0[b]	1320[a]
KNO_3	3000	100[a]	100[a]	3.2[b]
NH_4NO_3	3000	100[a]	100[a]	8.5[b]
$(NH_4)_2SO_4$	3000	100[a]	100[a]	132[b]
$(NH_4)_2CO_3$	3000	0[b]	0[b]	158[a]
Ammonia water	400	0[b]	0[b]	1325[a]
Control[3]	100[a]	100[a]	3.5[b]	

[1]Concentration of NH_4^+ (mg l^{-1}) was analysed by ion exchange chromatography.
[2]In each column values followed by the same letter are not significantly different according to the least significant difference test ($P = 0.05$). Each value is the mean of ten and six replicates for survival and concentration, respectively.
[3]Soil without chemicals was used as a control.

Table 7.2. Effects of soil type on generation of volatile NH_3 from urea and *P. noxius* survival in inoculated wood sections buried in the soil.

Soil source	pH	Survival of *P. noxius* (%)	Concentration of NH_4^+ (mg l^{-1})[1]
YMS3	6.1	100[a2]	3.3[c]
TCS	7.3	0[b]	1128[a]
SCS	5.6	100[a]	126[b]
SSS	7.9	0[b]	1832[a]
HLS	7.2	0[b]	1650[a]
TMS	6.5	100[a]	187[b]
KSS	9.2	0[b]	1358[a]
CHS	5.2	100[a]	252[b]

[1]Concentration of NH_4^+ (mg l^{-1}) was analysed by ion exchange chromatography; 3000 p.p.m. urea was added.
[2]In each column values followed by the same letter are not significantly different according to the least significant difference test ($P = 0.05$). Each value is the mean of ten and six replicates for survival and concentration, respectively.
[3]YMS, Yan-ming-shan; TCS, Taichung Harbour area; SCS, Hsin-chu City; SSS, Commercial sand for construction; HLS, Hualien coast; TMS, Taimalee, Taitung County; KSS, Kaohsiung County; CHS, Hsin-chu City.

Conclusion

Based on the numbers and dates of papers published and cited in forestry abstracts on six common root and butt rot fungi, *H. annosum*, *Armillaria* spp., *P. weirii*, *Inonotus tomentosus* (Fr.) Teng, *Phaeolus schweinitzii* (Fr.) Pat. and *P. noxius*, research on *P. noxius* has been initiated in more recent years and has the fewest papers published on it (Stenlid, 1998). However, it is believed that the economic losses caused by *P. noxius* in warmer climates are not proportional to

Table 7.3. Effects of soil matrix potential on generation of NH_3 from urea and *P. noxius* survival in inoculated wood sections.

Soil matrix potential (MPa)	Survival of *P. noxius* (%)	Concentration of NH_4^+ (mg l^{-1})[1]
–0.75	0[b2]	1125[a]
–0.50	0[b]	1231[a]
–0.30	0[b]	1025[a]
–0.15	0[b]	1102[a]
–0.025	100[a]	4.8[b]
Flooding	100[a]	3.8[b]

[1]Concentration of NH_4^+ (mg l^{-1}) was analysed by exchange chromatography; 3000 p.p.m. urea was added.
[2]In each column values followed by the same letter are not significantly different according to the least significant difference test ($P = 0.05$). Each value is the mean of ten and six replicates for survival and concentration, respectively.

the number of papers published compared with the other five root and butt rot fungi. It is remarkable that there were no published papers on the disease control of *P. noxius* brown root disease. Hopefully, more scientists will begin to study the impact, control and management of *P. noxius* brown root disease in the near future.

References

Angwin, P.A. (1985) Environmental factors influencing the rate of spread of *Phellinus weirii* (Murr.) Gilbertson in young-growth Douglas-fir. MS Thesis, Oregon State University, Corvallis, Oregon.

Ann, P.J. and Ko, W.H. (1992) Decline of longan tree: association with brown root rot caused by *Phellinus noxius*. *Plant Pathology Bulletin* 1, 19–25.

Ann, P.J., Lee, H.L. and Huang, T.C. (1999a) Brown root rot of 10 species of fruit trees caused by *Phellinus noxius* in Taiwan. *Plant Disease* 83, 746–750.

Ann, P.J., Lee, H.L. and Tsai, J.N. (1999b) Survey of brown root disease of fruit and ornamental trees caused by *Phellinus noxius* in Taiwan. *Plant Pathology Bulletin* 8, 51–60.

Anonymous (1974a) Root diseases. Part I: Detection and recognition. *Rubber Research Institute Malaya Plant Bulletin* 133, 111–120.

Anonymous (1974b) Root diseases. Part II: Control. *Rubber Research Institute Malaya Plant Bulletin* 134, 157–164.

Bolland, L. (1984) *Phellinus noxius*: cause of a significant root rot in Queensland hoop pine plantation. *Australian Forestry* 47, 2–10.

Chang, T.T. (1992) Decline of some forest trees associated with brown root rot caused by *Phellinus noxius*. *Plant Pathology Bulletin* 1, 90–95.

Chang, T.T. (1995a) A selective medium for *Phellinus noxius*. *European Journal of Forest Pathology* 25, 185–190.

Chang, T.T. (1995b) Decline of nine tree species associated with brown root rot caused by *Phellinus noxius* in Taiwan. *Plant Disease* 79, 962–965.

Chang, T.T. (1996) Survival of *Phellinus noxius* in soil and in the roots of dead host plants. *Phytopathology* 86, 272–276.

Chang, T.T. and Chang, R.J. (1999) Generation of volatile ammonia from urea fungicidal to *Phellinus noxius* in infested wood in soil under controlled conditions. *Plant Pathology* 48, 337–344.

Chang, T.T. and Yang, W.W. (1998) *Phellinus noxius* in Taiwan: distribution, host plants and the pH and texture of the rhizosphere soils of infected hosts. *Mycological Research* 102, 1085–1088.

Chang, T.T., Chiu, W.H. and Young, W.W. (1994) Selective nitrogen assimilation by *Phellinus noxius* and related fungi. *Plant Pathology Bulletin* 3, 230–233 (in Chinese).

Hodges, C.S. and Tenorio, J.A. (1984) Root disease of *Delonix regia* and associated tree species in the Mariana Islands caused by *Phellinus noxius*. *Plant Disease* 68, 334–336.

Kuhlman, E.G., Hodges, C.S. Jr and Froelich, R.C. (1976) *Minimizing losses to* Fomes annosus *in the southern United States*. USDA Forest Service Paper SE-151. USDA Forest Service, Oregon.

Larsen, M.J. and Cobb-Poulle, L.A. (1990) *Phellinus* (*Hymenochaetaceae*) a survey of the world taxa. *Synopsis Fungorum* 3, 1–206.

Nandris, D., Nicole, M. and Geiger, J.P. (1987) Root rot diseases of rubber trees. *Plant Disease* 71, 298–306.

Neil, P.E. (1986) A preliminary note on *Phellinus noxius* root rot of *Cordia alliodora* plantings in Vanuatu. *European Journal of Forest Pathology* 16, 274–280.

Rishbeth, J. (1950) Observation on the biology of *Fomes annosus*, with particular reference to East Anglian pine plantations. I. The outbreaks of disease and ecological status of the fungus. *Annals of Botany* 55, 365.

Shaw, C.G. III and Kile, G.A. (1991) *Armillaria root disease.* Agriculture Handbook No. 691. USDA Forest Service, Washington, DC.

Singh, S., Bola, L. and Kumar, J. (1980) Diseases of plantation trees in Fiji Islands I, Brown root rot of mahogany (*Swietenia macrophylla* King). *Indian Forestry* 106, 526–532.

Stenlid, J. (1998) Opening remarks. In: Delatour, C., Guillaumin, J.J., Lung-Escarment, B. and Marcais, B. (eds) *Root and Butt Rot of Forest Trees.* INRA, Paris, pp. 13–17.

Thies, W.G. (1984) Laminated root rot. The quest for control. *Journal of Forestry* 82, 345–356.

Thies, W.G., Nelson, E.E. and Zabowski, D. (1994) Removal of stumps from a *Phellinus weirii* infested site and fertilization affect mortality and growth of planted Douglas-fir. *Canadian Journal of Forest Research* 24, 234–239.

Tsai, Y.P. (1991) *List of Plant Diseases in Taiwan*, 3rd edn. Plant Protection Society of the Republic of China & Plant Phytopathological Society of the Republic of China, Taichung, Taiwan.

Turner, P.D. (1981) *Oil Palm Diseases and Disorders.* Oxford University Press, Kuala Lumpur, Malaysia.

Wallis, G.W. (1976) Phellinus (Poria) weirii *root rot. Detection and management proposals in Douglas-Fir stands.* Canadian Forest Technical Report 12. Canadian Forestry Service, Victoria, BC.

Whitney, R.D. (1988) *The Hidden Enemy.* Canadian Forestry Service, Ontario, Canada.

Production of Ligninolytic Enzymes by Species Assemblages of Tropical Higher Fungi from Ecuador

8

M.A. Ullah,[1] R. Camacho,[2] C.S. Evans[1] and J.N. Hedger[1]

[1]*Fungal Biotechnology Group, University of Westminster, 115 New Cavendish Street, London W1M 8JS, UK;* [2]*Facultad de Ciencias, Escuela Superior Politecnica de Chimborazo, Riobamba, Ecuador*

Introduction

No systematic screening studies have been made on lignin-degrading enzymes produced by higher fungi isolated from tropical moist forests. This is a surprising anomaly, since net primary production in moist tropical forests exceeds that of any other terrestrial biome, with a litter production of 10–35 t ha^{-1} year^{-1}, of which wood can constitute 74% (Swift *et al.*, 1976). Assuming an average lignin content of 20%, the annual input of wood lignin is thus 1.5–5.3 t lignin ha^{-1}. The decomposition constant, k value, for moist tropical forest litter is high, estimated by Swift *et al.* (1976) at ~6.0, underlining the rapid turnover of lignocellulose in the ecosystem. Reviews, such as, by Jordan (1985) that have emphasized that fungi and termites are responsible for most of this decomposition.

The mechanisms of lignin degradation by fungi involve a variety of enzymes, including polyphenol oxidase (laccase) and the peroxidative enzymes, such as lignin-peroxidase and manganese-dependent peroxidase (Kirk and Farrell, 1987; Thurston, 1994). Lignin is one of the three major polymers in wood cell walls, which provide strength and rigidity, the others being cellulose and hemicellulose.

White rot fungi are considered to cause degradation of all three polymers, either as simultaneous or as sequential rots, leaving some of the white hemicellulose and cellulose behind (Kirk *et al.*, 1978; Evans *et al.*, 1991). In contrast, brown rots are considered to selectively degrade the cellulose and hemicellulose components, possibly by non-enzymic peroxide oxidation and with oxalic

 Tropical Mycology, Vol. 1, *Macromycetes* (eds R. Watling, J.C. Frankland, A.M. Ainsworth, S. Isaac and C.H. Robinson)

acid–hydrogen peroxide, leaving the lignin behind as a modified brown-coloured polymer (Kirk and Fenn, 1982; Dutton *et al.*, 1993). White rot fungi predominate in moist tropical forest, especially members of the *Agaricales*, aphyllophoroid taxa and *Xylariaceae*, but brown rot fungi are almost completely absent, possibly because they co-evolved with the *Coniferae* in the boreal forest biome (Watling, 1982). No detailed studies on the ligninolytic enzymes produced by tropical wood, white rot fungi are known, or on the other major group of fungal decomposers in rainforests, the litter-degrading fungi. The readily observed bleaching of forest floor leaf litter, which is caused by the mycelia of such fungi, indicates that exploitation must also involve degradation of lignin. However, the physico-chemical environment of the small litter resources is very different from that of larger volumes of wood, and could influence the types of ligninases used to depolymerize them.

Accordingly, it was thought that a survey of the ligninolytic systems found in the species assemblages of fungi on these two resource types, wood and litter, would provide much needed data on the influence of resource selectivity on the exploitative enzymes produced by the fungi.

Methods

Study site

The fungal cultures used in the study, apart from *Trametes versicolor* Quélet (FPRL-28A supplied by CABI, Egham, UK), had been isolated from Rio Palenque, a 87 ha reserve located on the west side of the Cordillera in Ecuador some 65 km south of the Equator, at an altitude of 150 m. The flora has been fully described by Dodson and Gentry (1978). It consists of primary Pluvial Pacific Coastal forest, with a very high species diversity, and of disturbed forest rich in gap colonizing trees such as balsa (*Ochroma pyrimidale* (Cav : Lam.) Urb.) together with extensive clumps of bamboo (*Bambusa guadua* H. & B.). An account of the ecology of decomposer fungi in this zone of Ecuador is given in Hedger (1985) and of the Amazonian forest in Ecuador by Lodge and Cantrell (1995).

Survey of fruit bodies

The survey of the resource relations of higher fungi in Rio Palenque was carried out by random fruit body collection in September 1997. Over 100 collections were made. Two resource types were distinguished: wood and small litter. Wood included all parts of fallen and standing dead, or partially dead, trees, down to branches > 20 mm in diameter, while small litter included twigs < 20 mm in diameter, leaf litter and reproductive structures. The resource component on which each fruit body occurred and the state of decay immediately beneath it was noted, in order to infer the resource relationships of the mycelium (Rayner *et al.*,

1985; Frankland, 1992). This information was used to group the taxa into resource-based species assemblages, and was combined with an existing database obtained from similar surveys carried out previously in Rio Palenque, between 1981 and 1996, in which 793 collections of macrofungi were made (J.N. Hedger, London, unpublished data). The authors prefer the use of the term species assemblage to community, since the fruit body distribution could not provide implicit evidence that mycelia of these taxa were part of an interactive fungal community.

Isolation of mycelia

Isolations were carried out in the field in September 1997, using a portable Perspex isolation cabinet, which could be sterilized by swabbing out with alcohol. Portions of fruit bodies were inoculated on to pre-poured selective media in 2.5 cm diameter plastic Petri dishes, using a spirit lamp to flame the isolation tools, and sealed with Parafilm. For agarics, a spore-fall method was also used, in which a pileus was attached to the lid of a Petri dish with sterile Vaseline, left to deposit basidiospores overnight on to the medium and removed the next morning. Members of the *Sarcoscyphaceae* and *Xylariaceae* were isolated from ascospores discharged from portions of the ascomes stuck on the lids of Petri dishes containing media and incubated upside down. The media used were selective for basidiomycetes or ascomycetes and are described by Hedger (1982). After incubation at room temperature (20–29°C) for 24–36 h, in a sealed plastic bag to prevent entry of ants, the plates were checked from below for presence of clamp connections (basidiomycetes) and ascospore germination and hyphal morphology (ascomycetes). Inocula were then transferred in the laboratory to 3% malt agar (MA) or 2% potato dextrose agar (PDA) slants and grown at 26°C; plugs of these mycelial cultures were stored in crio-vials under sterile water at 20°C. The fungi were recovered by transferring inoculum on to the centre of Petri plates of the same medium, and incubated for 7 days at 26°C, before checking hyphae for purity. For screening studies, 4 mm diameter inocula marked out with a sterile cork borer were removed from edges of 3–7-day-old, 4–6 cm diameter colonies, growing on 3% MA or 2% PDA plates at 26°C.

Screening for ligninolytic enzymes

Ten taxa from each species assemblage were screened for ligninolytic enzyme production in liquid culture. The basis of the choice is described later. Measurements showed that, in cultures in which ligninolytic enzyme activity had been induced, these enzymes represented up to 90% of total protein in the culture filtrate. Activity measurements are presented as activity units per mg dry wt mycelial biomass, which was considered to be a more realistic comparative measure of ligninolytic activity of the isolates than specific activity of the protein fraction of the culture filtrates.

Laccase production

For laccase production, a liquid glucose–amino acid–salts medium was used (Fahraeus and Reinhammar, 1967). Conical flasks (50 ml) containing 25 ml of medium were inoculated in triplicate with 4 mm × 4 mm diameter mycelial plugs for each organism. The flasks were shaken at 26°C for 7 days, when 2,5-xylidine was added 18 h before harvesting to give a concentration of 2×10^{-4} M in the growth medium (Fahraeus and Reinhammar, 1967). Mycelium was then separated from the medium by filtration and samples of 5 – 20 μl culture filtrate were removed and added to 1.0 ml of catechol (10 mM) in 100 mM sodium acetate buffer at pH 5, and laccase activity measured by increase in absorbance at 440 nm (Evans and Palmer, 1983). One unit of laccase was defined as causing a change in absorbance of 1.0 ml^{-1} min^{-1} at 25°C, each value being the mean of three replicate cultures.

Manganese peroxidase (MnP) production

For MnP production, static cultures were prepared in 300 ml medical flats containing 50 ml of low-nitrogen medium adapted from Glenn and Gold (1985). The medium contained glucose (20 g l^{-1}), ammonium tartrate (1.2 mM), mineral salts and vitamins. The mineral salts (g l^{-1}) included 0.01 $CaCl_2$, 0.01 $FeSO_4 . 7H_2O$, 0.001 $MnSO_4 .4H_2O$, 0.001 $ZnSO_4 .7H_2O$, 0.002 $CuSO_4$. The vitamin solution (mg l^{-1}) included 27.5 adenine, 0.05 thiamine HCl, 1000 *I*- inositol, 100 pyridoxine-HCl, 100 nicotinic acid, 100 sodium pantothenate, 100 *p*-aminobenzoic acid, 100 riboflavine, 30 biotin, 10 folic acid, 10 cyanocobalamin. The flasks were incubated statically for 14 days at 26°C, and MnP activity was measured by the transformation of 100 μM of 2,6-dimethoxyphenol (DMP) to coerulignone (Martinez *et al.*, 1996). The reaction mixture included 100 μM $MnSO_4$ in 100 mM sodium tartrate buffer at pH 5.0 and 100 μM H_2O_2 with 5–20 μl enzyme extract to be assayed. Absorbance was measured at 238 nm with a molar extinction coefficient of 6500 M^{-1} cm^{-1} and one unit of MnP was equivalent to one mole of coerulignone formed under reaction conditions (c = A/E), each value being the mean of three replicate cultures.

Results

Species assemblages of fungi in Rio Palenque

The analysis of the 1997 data and those of previous years showed that the two major resource-related species assemblages of saprobic fungi in Rio Palenque, on small litter and on wood, together accounted for over 90% of the collections. Fungi on soil and arthropods together made up only around 10% of the collections. No obviously mycorrhizal taxa were found, except for one unidentified species of *Boletus* in 1982.

The species assemblages on small litter and wood, respectively, accounted for approximately 30% and 60% of the macrofungi surveyed. A number of taxa were common to both of these assemblages, especially the species found both on small litter and on the smaller woody resources, branches 20–40 mm diameter. However the surveys showed that this species overlap was only ~20% of the total, 80% of the taxa fruiting exclusively on either wood or small litter. There was no overlap at all between the species fruiting on small litter and those present on the larger woody resources, branches > 50 mm diameter and tree trunks. Lodge (1997) also found, in a resource survey of rainforest fungi in Puerto Rico, that only 173 (25%) of the 705 decomposer fungi she recorded occurred on two resource types, whereas 493 taxa (70%) occurred on only one resource type.

Ten isolates were selected from each species assemblage for further study of ligninolytic enzymes. The basis of selection was to attempt to sample a broad spectrum of mycelial behaviour from within each assemblage, including different decomposition strategies and differences in component and taxon selectivity. The fungal taxa selected are shown in Table 8.1, arranged with respect to their species assemblages. Also included in this table is a field assessment of the state of decomposition of the resource on which the fruit bodies were found, although whether the mycelia of these taxa had caused this decomposition could not be proved.

The mycelia of all fungi on small litter caused bleaching or white rot of the litter material, implying that they were capable of simultaneous degradation of lignin, hemicellulose and cellulose. Much variation in component and taxon selectivity operated within this assemblage, related to the mixture of available resources (leaves, twigs, fruits, seeds and other plant debris). Some taxa were foliicolous, almost entirely restricted to leaves. The genus *Marasmius* exemplified this strategy, and many species appeared to be component- and taxon-selective. The mycelia of these species could be easily identified as discrete patches of bleaching, always on intact leaves in the upper litter layer on the soil surface, and also on leaves trapped in the understorey shrubs. However the identities of the leaves on which each *Marasmius* species occurred were difficult to establish. *Marasmius cladophyllus*, which was readily observed to be taxon-selective for leaves and petioles of *B. guadua*, was isolated for further study. *Favolaschia cinnabarina* was also chosen, because it was obviously component-taxon selective for petioles of palm and *Heliconia* species. Other taxa isolated, such as *Collybia bakeri*, *Xeromphalina tenuipes*, and *Lepiota lactea* had non-component-selective mycelia, which advanced as a front colonizing all litter components. Four aphyllophoroid taxa, *Polyporus obolus*, *Polyporus lepreurii*, *Trametes ochracea* and *Coltricia spathulata* were also chosen, since they were clearly component-selective decomposers, occurring only on small twigs and branches at soil level, which they bound together by pseudosclerotial plates (psps; Ainsworth and

Table 8.1. Fungal cultures selected for ligninolytic enzymes studies from two species assemblages studied in Rio Palenque forest.

Small litter assemblage	Family	Rot type
Collybia bakeri Dennis	*Tricholomataceae*	B
Lepiota lactea Murrill	*Agaricaceae*	B
Marasmius cladophyllus Berk.	*Tricholomataceae*	B
Xeromphalina tenuipes (Schwein.) A.H.Sm.	*Tricholomataceae*	B
Favolaschia cinnabarina (Berk. & M.A Curtis) Pat.	*Tricholomataceae*	WR
Nothopanus eugrammus (Mont.) Singer	*Tricholomataceae*	WR
Coltricia spathulata (Hooker) Murrill	*Hymenochaetaceae*	WR
Polyporus obolus Ellis. & T. Macbr.	*Polyporaceae*	WR
Polyporus lepreurii Mont.	*Polyporaceae*	WR
Trametes ochracea (Pers.) Gilb. & Ryvarden	*Coriolaceae*	WR
Lignicolous assemblage		
Cookeina tricholoma (Mont.) Kuntze	*Sarcoscyphaceae*	WR
Irpex maximus Mont.	*Steccherinaceae*	WR
Lentinus crinitus (L.) Fr.	*Lentinaceae*	WR
Lentinus strigosus (Schwein.) Fr.	*Lentinaceae*	WR
Stereum insigne Bres.	*Stereaceae*	WR
Trametes corrugata (Pers.) Bres.	*Coriolaceae*	WR
Xylaria guyanensis (Mont.) Mont.	*Xylariaceae*	WR
Psathyrella albocapitata Dennis.	*Coprinaceae*	WR
Filoboletus gracilis (Klotzsch : Berk.) Singer	*Tricholomataceae*	SR
Tomentella lilacinogrisea Wakef.	*Thelephoraceae*	?WR

B, bleaching of leaf litter by mycelium; WR, white rot of wood; SR, soft rot of wood.

Rayner, 1990; Hedger *et al.*, 1993). In the case of *P. lepreurii*, 2–4 mm diameter negatively gravitropic rhizomorphs were also produced. It is probable that this fungus was also taxon-selective, since its distribution in the forest was very disjunct. The sessile, tricholomatoid agaric *Nothopanus eugrammus* was also studied, since it appeared to have little component or taxon selectivity. It was most commonly found fruiting on twigs but also occurred on larger diameter tree branches.

In the lignicolous species assemblage, the wood below the fruit bodies was usually partially or completely bleached, sometimes with discoloration by pigments and psps, typical of fungal white rot attack. Association between fruit bodies and the mycelia in the decomposed wood below was investigated by cutting into the wood, using indicators such as relative wood bleaching, pigment production, odour and psps to determine that the mycelium in the wood had given rise to the fruit body. These decomposers could be divided on component selectivity and mode of attack on the wood. On fallen trunks, on the drier upper portions of the logs, there were large fruitings of many obvious decomposers. Examples of basidiomycetes selected for study were *Irpex maximus*, *Lentinus crinitus*, *Lentinus strigosus* and *Psathyrella albocapitata*, all of which were associated with softened, moist, bleached wood. The ascomycetes *Xylaria guyanensis* and *Cookeina tricholoma* were also included in the study. Their ascomes, and those of other *Xylariaceae* and *Sarcoscyphaceae,* were associated with bleached wood which was hard and dry, with surface and internal psps present. On the wetter undersides of the logs the dominant decomposers were resupinate aphyllophoroid taxa, the example selected being *Stereum insigne*. *Tomentella lilacinogrisea* was also identified growing in this situation and its fruit bodies seemed to be associated with a white rot mycelium. However, A.M. Ainsworth (Slough, UK, personal communication) opines that this fungus is mycorrhizal, so that the association of mycelium with fruit bodies may have been incorrect in this instance. Its ecological status in Rio Palenque thus remains uncertain. Finally, *Filoboletus gracilis* was selected as a representative of soft rot attack. This fungus was always found on wood semi-immersed in small streams in the forest, and associated with grey, water-soaked, highly softened wood.

Production of ligninases

Figures 8.1 and 8.2 show the results of screening the twenty isolates for MnP and laccase activity. The results are grouped according to species assemblage, rather than taxonomic affinities.

Manganese peroxidase (MnP) activity in the species assemblages

The results of MnP assays of the culture filtrates are shown in Fig. 8.1. Although the results were variable, average values for the enzyme units of the two species assemblages showed an opposite trend to the laccase units, being significantly lower in the lignicolous assemblage, 4.3 units, than in the small litter assemblage, 12.5 units (t-statistics = 1.777, $P = 0.05$). Five of the small litter assemblage taxa had MnP titre values > 10 units in the culture filtrates (*C. bakeri*, *M. cladophyllus*, *P. obolus*, *P. lepreurii* and *T. ochracea*), while only one taxon in the lignicolous assemblage, *I. maximus*, had a value > 10 units. However, both species assemblages had three isolates showing zero or negligible MnP activity, *L. lactea*, *X. tenuipes* and *N. eugrammus* in the small litter assemblage and *L. crinitus*, *L. strigosus* and *X. guyanensis* in the lignicolous assemblage. The comparative figure for *T. versicolor* was 14.6 units.

Laccase activity in the species assemblages

The laccase titres (Fig. 8.2) of fungi from the lignicolous species assemblage showed a significantly higher average level of laccase, 284.8 units (t-statistics = 2.16, $P = 0.05$), than the mean for the taxa from the small litter species

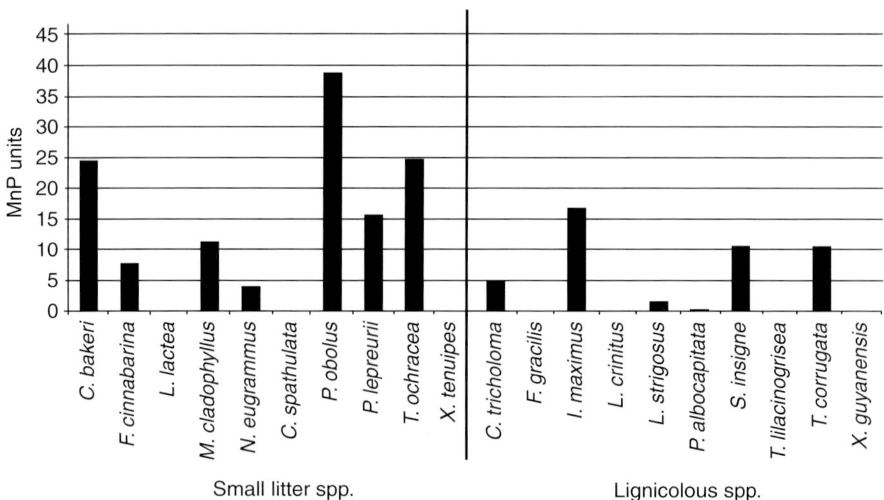

Fig. 8.1. Manganese peroxidase (MnP) activity of culture filtrates. One unit of MnP was equal to one mole of coerulignone formed under reaction conditions, using 100 μM of 2,6-dimethoxyphenol (DMP), each value being the mean of three replicate cultures.

assemblage, 6.4 units. Two of the small litter assemblage taxa had no detectable laccase in the culture filtrates (*N. eugrammus* and *T. ochracea*), and in *M. cladophyllus*, *C. bakeri* and *X. tenuipes* activity was barely detectable. Although the other five litter taxa had some laccase activity in their culture filtrates, the titre was much lower than that of the lignicolous species, the highest being found in the filtrates from *L. lactea* (18.2 units). In contrast, eight of the lignicolous species had a titre above 10 units, five had activities in excess of 100 units and *I. maximus* and *L. crinitus* had activities of more than 1000 units. Of particular interest were the two cultures from the lignicolous species assemblage which had titres lower than 10 units, *C. tricholoma* (5.4), and *X. guyanensis* (0), indicating there was little, if any, laccase expression by these fungi. The comparative laccase titre for *T. versicolor* was 1802 units.

The results in Figs 8.1 and 8.2 also show that two taxa, *X. guyanensis* and *X. tenuipes*, had zero, or near zero, values for both ligninolytic enzymes in their culture filtrates. Both taxa grew well on the culture medium and field data showed that both caused very active *in vivo* degradation of lignocellulose, *X. guyanensis* of wood and *X. tenuipes* of litter.

Discussion

The production of ligninolytic enzymes by these tropical fungi can be compared with the resource selectivity of their mycelia in the field. The clearest association was between high laccase levels and the wood species assemblage. Laccase is considered to play a substantial role in the degradation of wood cell walls by

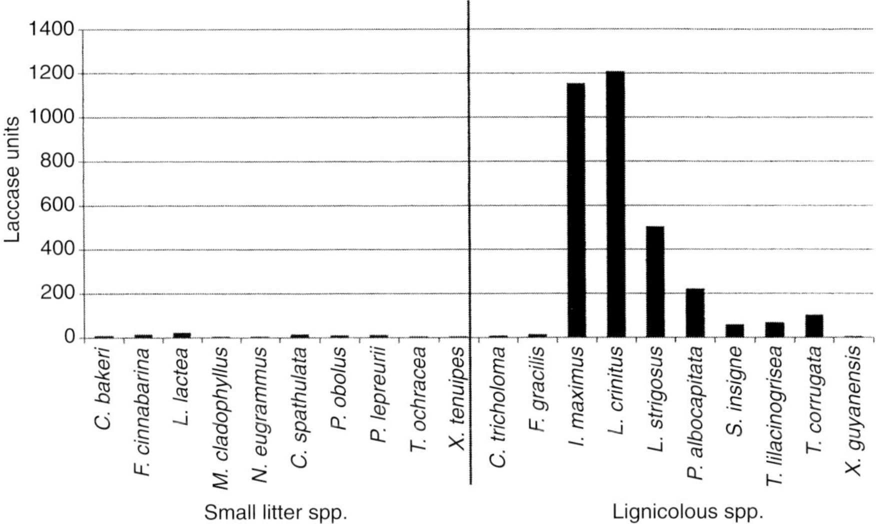

Fig. 8.2. Laccase activity in culture filtrates. One unit of laccase was defined as causing a change in absorbance of 1.0 min^{-1} ml^{-1} at 25°C, each value being the mean of three replicate cultures.

first attacking phenolic groups on lignin polymers, so generating free radicals for depolymerization reactions to occur, ultimately leading to the exposure of cellulose and hemicellulose to attack by removal of lignins from the s1 and s2 layers of the cell wall (Evans *et al.*, 1994). The data confirmed the importance of the enzyme in these tropical lignicolous fungi. The low laccase levels found in the litter species are less easy to understand. Attack on leaf litter and small branches by these fungi resulted in bleaching, usually followed by conversion to powder, indicating rapid simultaneous removal of lignin, cellulose and hemicellulose. It is true that the proportion of secondarily thickened wall material in the small litter resource must be lower than in wood (Jordan, 1985; Rayner and Boddy, 1988) and the few studies on the decomposition of the two resource types in tropical forest showed that the *k* values of wood are much lower than those of small litter (Swift *et al.*, 1976). However, it might be expected that laccase production by these litter decomposer fungi would be necessary for attack on the significant amount of lignin present, especially in the small twigs and branches. However, even the litter fungi that were component-selective for lignin-rich twigs and leaf petioles, the four aphyllophoroid taxa and *F. cinnabarina*, had a low laccase titre.

On the other hand, the higher MnP titre of the small litter species assemblage raises the possibility that this ligninase is more important to fungi in resources with a lower lignin content or a more diffuse distribution of lignified tissue. Culture conditions in the laboratory were optimized for fungal production of MnP and laccase to demonstrate the maximal capability of the organism for enzyme production. Under these conditions of optimized production more laccase was produced in the wood assemblage species and more MnP in the litter assemblage. Ligninolytic enzyme activity depends on the substrate, the co-reactants (H_2O_2/O_2), and the pH of the environment. Both laccase and MnP react with phenolic lignin groups, although under some conditions they can attack non-phenolic moieties when appropriate mediators are present (Leontievsky *et al.*, 1997). Wood contains a larger proportion of lignin (~15–20%) than is present in litter (~5–10%) (Rayner and Boddy, 1988). More enzyme would therefore be required to degrade the wood lignin than the small-litter lignin and enable the fungus to gain access to the cellulose and hemicellulose in the resource. Increasing production of ligninases to meet this need may be more readily met by increased synthesis of laccase than MnP. In temperate species of fungi, induction of neutral laccases can be stimulated by addition of Cu^{2+} as in *Pleurotus ostreatus*, and acidic laccases by lignin mimics, such as 2,5-xylidene in *T. versicolor* (Palmieri *et al.*, 2000). The availability of oxygen for laccase activity is unlikely to be a metabolic block in the two resources, although it is probably lower in the wood resource (Hintikka, 1982), but hydrogen peroxide, a co-reactant for MnP, may be more available to the species of the litter assemblage due to the higher O_2 and more effective hydrogen peroxide production systems.

Perhaps the most likely reason for greater laccase production by wood decomposers and MnP production by litter decomposers is the environmental pH. The pH in wood is around 4, while the usual pH of litter in tropical forest is ~5–6 (Swift *et al.*, 1976; Rayner and Boddy, 1988). In the laboratory, the optimum pH for enzyme activity for laccase is 4.0–5.0, with most phenolic substrates, and for MnP pH 5–6. It is therefore possible that production of laccase by the tropical lignicolous fungi is an adaptation to a surrounding pH more suitable for optimizing its activity, while MnP is produced by tropical litter fungi in an environment where pH is higher. If enzymes are produced that have pH optima close to the environment in which the mycelium is growing, then less enzyme is required, thus reducing the amount of protein synthesis required by the fungus, an important factor in the nitrogen-poor environment of wood and litter in tropical forest, where competition for available nutrients is severe (Lodge, 1993).

The search for ligninase enzyme systems has concentrated on fungi almost wholly selected from the temperate, wood decomposer community and especially on just a few taxa, such as *T. versicolor*. This study of tropical higher fungi indicates that the ligninase systems produced by a fungal taxon are related to the type of resource exploited by the mycelium, so that wider community-focused studies on the fungi of wood, litter and soil resources may reveal different balances of laccase, MnP and perhaps novel ligninase enzyme systems than current models.

Acknowledgements

We thank Calway and Piedad Dodson for hospitality in Rio Palenque, the Departamento de Ciencias Biologicas, Escuela Superior Politecnica de Chimborazo (ESPOCH), Riobamba, Ecuador, Fran Fox, Ricardo Viteri and Liam Evans for help in the field and laboratory studies, and the University of Westminster for a Quintin Hogg Scholarship for M.U. We are also grateful to Lynne Boddy, Juliet Frankland, Martyn Ainsworth, Sue Isaac, Clare Robinson and Habiba Gitay for help during the preparation of the text.

References

Ainsworth, A.M. and Rayner, A.D.M. (1990) Aerial transfer by *Hymenochaete corrugata* between hazel stems and other trees. *Mycological Research* 94, 263–266.

Dodson, C.H. and Gentry, A.H. (1978) Flora of Rio Palenque Science Center, Los Rios, Ecuador. *Selbyana* 4, 1–628.

Dutton, M.V., Evans, C.S., Atkey, P.T. and Wood, D.A. (1993) Oxalate production by basidiomycetes. *Applied Microbiology and Biotechnology* 39, 5–10.

Evans, C.S. and Palmer, J.M. (1983) Ligninolytic activity of *Coriolus versicolor*. *Journal of General Microbiology* 129, 2103–2108.

Evans, C.S., Gallagher, I.M., Atkey, P.T. and Wood, D.A. (1991) Localisation of degradative enzymes in white rot decay of lignocellulose. *Biodegradation* 2, 25–31.

Evans, C.S., Dutton, M.V., Guillen, F. and Veness, R.G. (1994) Enzymes and small molecular mass agents involved in lignocellulose degradation. *FEMS Microbiology Reviews* 13, 235–240.

Fahraeus, G. and Reinhammar, B. (1967) Large scale production and purification of laccase from the fungus *Polyporus versicolor* and some properties of laccase. *Acta Chemica Scandica* 21, 2367–2378.

Frankland, J.C. (1992) Mechanisms in fungal successions. In: Carroll, G.C. and Wicklow, D.T. (eds) *The Fungal Community. Its Organisation and Role in the Ecosystem.* Marcel Dekker, New York, pp. 383–401.

Glenn, J.K. and Gold, M.H. (1985) Purification and characterization of an extracellular Mn-II-dependent peroxidase from the lignin-degrading basidiomycete, *Phanerochaete chrysosporium. Archives of Biochemistry and Biophysics* 242, 329–341.

Hedger, J.N. (1982) Isolation of cellulose and lignin decomposing fungi from wood. In: Primrose, S.B. and Wardlaw, A.C. (eds) *Source Book of Experiments for the Teaching of Microbiology*. Academic Press, London, pp. 608–616.

Hedger, J.N. (1985) Tropical agarics: resource relations and fruiting periodicity. In: Moore, D., Casselton, L.A., Wood, D.A. and Frankland, J.C. (eds) *Developmental Biology of Higher Fungi.* Cambridge University Press, Cambridge, pp. 41–86.

Hedger, J.N., Lewis, P. and Gitay, H. (1993) Litter trapping by fungi in moist tropical forest. In: Isaac, S. Frankland, J.C., Watling, R. and Whalley, A.J.S. (eds) *Aspects of Tropical Mycology*. Cambridge University Press, Cambridge, pp. 15–35.

Hintikka, V. (1982) The colonisation of litter and wood by basidiomycetes in Finnish forests. In: Frankland, J.C., Hedger, J.N. and Swift, M.J (eds) *Decomposer Basidiomycetes, their Biology and Ecology*. Cambridge University Press, Cambridge, pp. 227–239.

Jordan, C.F. (1985) *Nutrient Cycling in Tropical Forest Ecosystems.* John Wiley & Sons, New York.

Kirk, T.K. and Farrell, R.T. (1987) Enzymatic 'combustion' – the microbial degradation of lignin. *Annual Review of Microbiology* 41, 465–505.

Kirk, T.K. and Fenn, P. (1982) Formation and action of the ligninolytic system in basidiomycetes. In: Frankland, J.C., Hedger, J.N. and Swift, M.J. (eds) *Decomposer Basidiomycetes, their Biology and Ecology*. Cambridge University Press, Cambridge, pp. 67–89.

Kirk, T.K., Schulze, E., Connors, W.J., Lorenz, L.F. and Zeikus, J.G. (1978) Influence of culture parameters on lignin metabolism by *Phanerochaete chrysosporium. Archives in Microbiology* 117, 277–285.

Leontievsky, A.A., Vares, T., Lankinen, P., Shergill, J.K., Pozdnyyakova, N.N., Myasoedova, N.M., Kalkinen, N., Golovleva, L.A., Cammack, R., Thurston, C.F. and Hatakka, A. (1997) Blue and yellow laccases of ligninolytic fungi. *FEMS Microbiology Letters* 156, 9–14.

Lodge, D.J. (1993) Nutrient cycling by fungi in moist tropical forest. In: Isaac, S., Frankland, J.C., Watling, R. and Whalley, A.J.S. (eds) *Aspects of Tropical Mycology*. Cambridge University Press, Cambridge, pp. 37–57.

Lodge, D.J. (1997) Factors related to the diversity of decomposer fungi in tropical forests. *Biodiversity and Conservation* 6, 681–688.

Lodge, D.J. and Cantrell, S. (1995) Fungal communities in wet tropics: variation in time and space. *Canadian Journal of Botany* 73 (suppl.), S.1391–1398.

Martinez, M.J., Ruiz-Duenas, F., Guillen, F. and Martinez, A.T. (1996) Purification and catalytic properties of two Manganese peroxidase isoenzymes from *Pleurotus eryngii. European Journal of Biochemistry* 237, 424–432.

Palmieri, G., Giardina, P., Bianco, C., Fontanella, B. and Saania, G. (2000) Copper induction of laccase isoenzymes in the ligninolytic fungus, *Pleurotus ostreatus. Applied and Environmental Microbiology* 66, 920–924.

Rayner, A.D.M. and Boddy, L. (1988) *Fungal Decomposition of Wood: its Biology and Ecology*. John Wiley, New York.

Rayner, A.D.M., Watling, R. and Frankland, J.C. (1985) Resource relations – an overview. In: Moore, D., Casselton, S., Wood, D.A. and Frankland, J.C. (eds) *Developmental Biology of Higher Fungi.* Cambridge University Press, Cambridge, pp. 1–40.

Swift, M.J., Heal, O.W. and Anderson, J.M. (1976) *Decomposition in Terrestrial Ecosystems.* Blackwell Scientific Publications, Oxford.

Thurston, C.F. (1994) The structure and function of fungal laccases. *Microbiology* 140, 19–26.

Watling, R. (1982) Taxonomic status and ecological identity in the basidiomycetes. In: Frankland, J.C., Hedger, J.N. and Swift, M.J. (eds) *Decomposer Basidiomycetes, their Biology and Ecology*. Cambridge University Press, Cambridge, pp. 67–89.

Laboratory Studies with *Leucoagaricus* and Attine Ants 9

P.J. Fisher[1] and D.J. Stradling[2]

[1]*School of Biological Sciences, University of Portsmouth, King Henry Building, King Henry 1st Street, Portsmouth PO1 2DY, UK;* [2]*Hatherly Laboratories, Department of Biological Sciences, University of Exeter, Exeter EX4 4PS, UK*

Formation of *Leucoagaricus* Basidiomes in a Live Nest of *Atta cephalotes*

Möller (1893) was the first to provide detailed descriptions of fungi isolated from ant fungus gardens. In his study the fungi were isolated from the nests of various attine species, including members of the genera *Acromyrmex*, *Apterostigma* and *Cyphomyrmex*, and he described and drew the development of agaricaceous basidiomes obtained from the surfaces of leaf-cutting ants' nests in the field. Accurate diagrams of gongylidia were made. These are the typical swollen hyphal tips of the vegetative mycelium in ants' nests, which developed after inoculation of nutrient agar with fragments of basidiome removed from inner non-fertile tissues. He also successfully grew vegetative mycelium from basidiospores which later formed gongylidia. Möller named the species *Rozites gongylophorus* A. Möller. Weber (1957, 1966), using the fungi isolated from nests of the primitive attines *Cyphomermex costatus* Mann and *Myrmicocrypta buenzli* Borgmeier, and Hervey *et al.* (1977), using the fungus from the ant species *Apterostigma auriculatum* Wheeler, obtained basidiomes in culture, but in no case did they report gongylidia-bearing mycelium. According to Heim (1957) and Singer (1975) both Weber's and Hervey's agarics were specifically identical and congeneric with *Leucoagaricus* and should be known as *Leucoagaricus gongylophorus* (A. Möller) Singer. The combination *L. gongylophorus* was formally proposed by Singer (1986). More recently Muchovej *et al.* (1991) have described *Leucoagaricus weberi* J. Muchovej, Della

Lucia and R. Muchovej as a new species and the first agaric found growing in a living nest of *Atta sexdens* L.

This chapter describes the successive formation of four separate basidiome aggregates over a period of 9 weeks in a living nest of the tropical lowland leaf-cutter ant *Atta cephalotes* L.

Colonies of *A. cephalotes* with queens were collected in Trinidad during the months of July and August in 1980, 1986 and 1991 and subsequently maintained in the Department of Biological Sciences at the University of Exeter, UK. They were about 1 year old, with a single fungus garden *c.* 10 cm diameter. In addition to the above, five young *A. cephalotes* colonies were collected by R. Jackman at Gilpin Trace, Roxborough, Tobago, during November 1989 and delivered to the laboratory at Exeter 3 days later, but the fungus gardens were destroyed in transit. To regenerate the colonies, each surviving queen with her workers and brood were placed on a moist sponge cloth in a plastic container and supplied with approximately 20 g of mature fungus garden taken from a laboratory colony of *A. cephalotes* at Exeter, collected in July 1986 in the Caura Valley, Trinidad.

All five of these colonies survived, successfully adopting the new fungus gardens. Cultures were kept at 25–26°C and relative humidity 70–80% in a 12:12 photoperiod. The nests were housed under Perspex containers, isolated on separate boards supported by legs, resting in paraffin moats to prevent ant escape. Each colony was provided with a dish of water and a daily supply of a variety of plant material, principally privet leaves (*Ligustrum ovalifolium* Hassk.), grapefruit skins (*Citrus paradisi* MacFad.), orange skins (*Citrus aurantium* L.) and cultivated rose leaves and petals (*Rosa* sp. hybrid tea, see Fig. 9.1). Expansion of the colonies was normally restricted to 15 cm × 15 cm × 20 cm by their covers, but one colony was eventually rehoused in a larger container 20 cm × 20 cm × 30 cm where the ants ultimately expanded the fungus garden to occupy 75% of the available space with an estimated fresh weight of 1.75 kg.

Fig. 9.1. Ants carrying rose petals to their nest.

In June 1993 an irregular structure with the appearance of an incipient basidiome began to develop in this enlarged colony (Fig. 9.2). Other subsequent aggregations of new basidiomes developed at 2–3 week intervals, usually just above the central line of the fungus garden until after about 10 weeks the process terminated (Fig. 9.3). The addition of inert markers indicated that the turnover time for plant material was about 5–7 weeks, so the basidiomes gradually moved downwards during their formation as new fungus garden was added at the top of the colony and old garden was removed from below by the ants. The ants appeared to feed their larvae preferentially on developing basidiome tissue, perhaps because the fungus was present there in a concentrated form, similar to that of the gongylidia in the fungus garden, which are the principal larval food.

The basidiomes were hollowed out by the ants from the inside (Fig. 9.4), with the result that most of the hymenial tissues were destroyed before they could ripen to produce basidiospores. Some basidiome aggregates were therefore removed at an appropriate stage. Samples of teased tissue taken from the inside of the stipe showed a structure traversed by gloeoplerous hyphae with oleaginous contents which stained deeply with lactofuchsin (Fig. 9.5). Electron micrographs (longitudinal sections) revealed dolipore septa similar to those in the mycelium found in the fungus garden, which furnished additional proof that the structures were basidiomes. One basidiome was sectioned (Fig. 9.6). Squash preparations of hymenial tissue yielded four-spored basidia, 22–32 × 8.5–

Fig. 9.2. Developing basidiomes of *Leucoagaricus gongylophorus* on a fungus garden in a live nest of *Atta cephalotes*. Scale bar = 4 cm.

Fig. 9.3. Some stages of development of basidiomes removed from a nest of *Atta cephalotes*. Scale bar = 2 cm.

14 μm and individual basidiospores, 5.7–7.7 × 4.5–5 μm; cheilocystidia, 24–44 × 5.5–8.5 μm, taken from a gill edge, were also present. The strongly dextrinoid spore wall, the lack of apical differentiation into a germ pore or similar structure and the absence of clamp connections are all features typical of *Leucoagaricus*.

Attempts to germinate immature basidiospores failed and although a number of basidiomes were placed on moist cotton wool in a damp chamber, the basidiomes failed to expand to yield mature basidiospores from which gongylidia-forming mycelium could be cultured. To relate the basidiome tissue to the mycelial form of the fungus, ten aggregates of cheilocystidial elements, each aggregate measuring ≤ 1 mm, were then dissected from the hymenium under sterile conditions and plated on to potato dextrose agar (PDA) medium. Thirty fragments of stipe tissue were similarly dissected and incubated on PDA at 25°C. After about 1 week all the incubated fragments showed new mycelial growth and approximately 15 days later typical aggregates of gongylidia (Fig. 9.7) formed at the edges of these colonies. All colonies obtained from fragments of the basidiomes looked and behaved identically to those obtained from the fungus garden.

It is also noteworthy that ants secrete antibiotic substances which suppress the growth of many microorganisms (Maschwitz *et al.*, 1970). One such substance, myrmicacin, $C_{10}H_{20}O_3$ (Schildknecht and Koob, 1971), inhibits the growth of a large number of fungi, including typical soil organisms such as *Penicillium*, and plant epiphytes such as *Cladosporium* and *Alternaria*. This

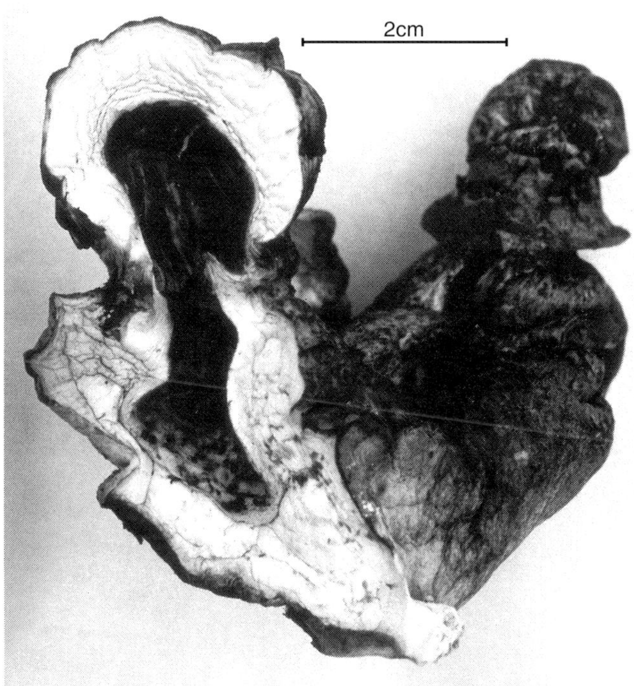

Fig. 9.4. A basidiome which has been hollowed out by *Atta cephalotes*. Scale bar = 2 cm.

indicates that the total mass of basidiome tissue (± 200 g) must have grown from the cultivated fungus since an invading fungus was likely to have been suppressed.

This work therefore strongly suggests a connection between the basidiomes and the mycelium of the fungus garden, although it would have been preferable to obtain further proof from germinating basidiospores. According to Muchovej *et al.* (1991), their description of *L. weberi* in a nest of *A. sexdens* was the first record of any agaricoid fungus in an active nest in which ants were still cultivating the fungus garden. The work at Exeter described a second such event but in a nest of *A. cephalotes.*

The authors' successful method for isolating mycelium from the basidiome tissue to obtain gongylidia reflects Möller's work, but his original herbarium material appears to have been lost. It is regrettable that, except for Möller, none of the other workers reported gongylidia formed by the vegetative mycelium obtained from basidiomes, a feature common to all fungi isolated from true attine fungus gardens. What induces the formation of teleomorphs in a fungus garden remains unknown. In the absence of clamp connections on the mycelium bearing the

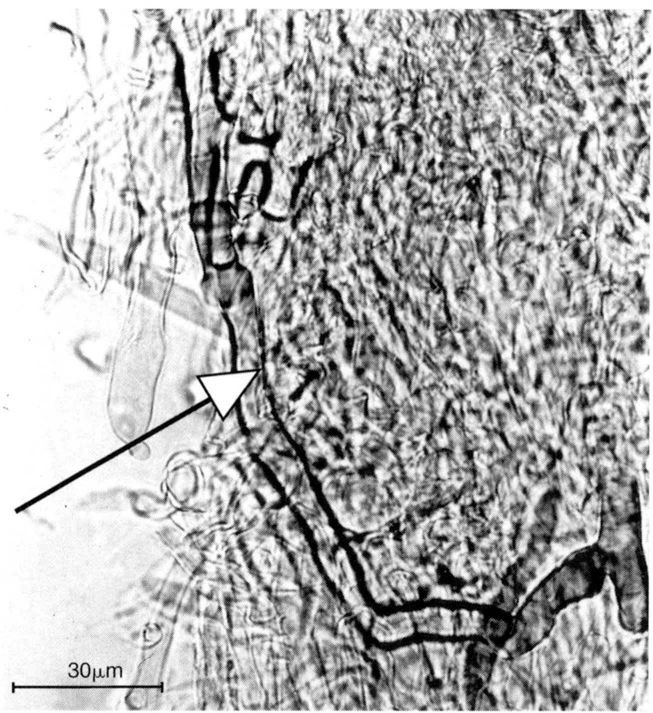

Fig. 9.5. Teased tissue taken from the inside of a stipe showing gloeoplerous hyphae (arrowed). Scale bar = 30 μm.

gongylidia, and observations on the cytological state of the mycelium, it is not known whether the mycelium in the nest is normally monokaryotic or dikaryotic. However, it seems unlikely that after nests had been kept under identical conditions at the Exeter laboratory for many years without apparently forming the teleomorph, sudden cytological changes in the mycelium in the nest should result in formation of basidiomes. It is more likely that some subtle change in the state of the nest had occurred, such as a failure of the ants to remove teleomorph initials for feeding their larvae, thus allowing the basidiomes to grow into more mature structures. This was supported by the fact that the queens of such nests, and consequently the whole colony, went into decline and in some cases died some weeks after the last basidiome had formed. Thus the colonies had probably entered the final ageing process when the ants allowed the basidiomes to form. Communications within an ant's nest must depend heavily on complex chemically induced, feedback mechanisms. Thus, when the queen stops laying eggs the absence of young larvae and the declining need to supply fungus food are probably communicated to the workers and directly affects their fungus-collecting behaviour. Since basidiome tissue is tough and probably more difficult to collect than gongylidia, the former is more likely to be neglected as the need for larval food declines.

Fig. 9.6. Freshly sectioned developing basidiome. Scale bar = 2.5 cm.

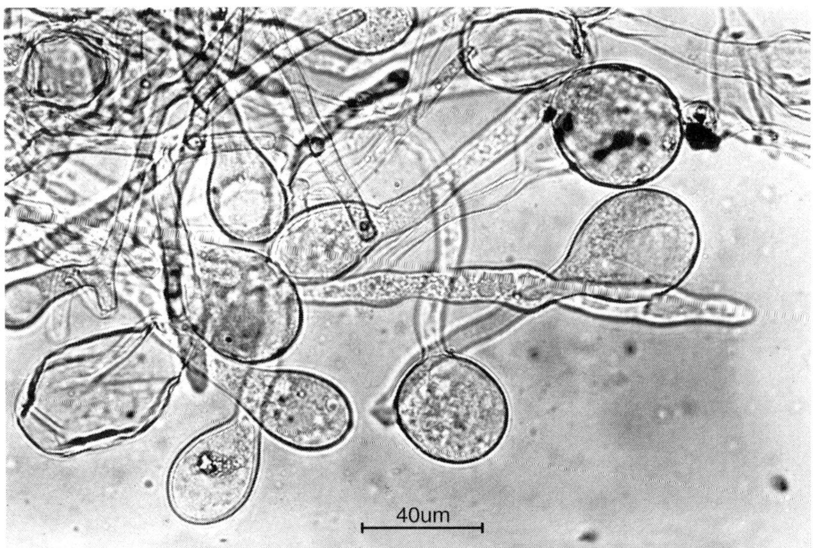

Fig. 9.7. Club-shaped gongylidia, typical of attine fungus gardens and the principal food of ant larvae. Scale bar = 40 μm.

An Assessment of Microfungi in the Fungus Gardens of *Atta cephalotes*

The ant fungus garden is traditionally considered to be a fungal monoculture on a vegetable substratum derived from a wide variety of plant species. The ants actively inoculate the substratum and maintain the culture by various physical and chemical means (Maschwitz *et al.*, 1970; Schildknecht and Koob, 1971). Since Belt (1874) realized that attine ants cultivated a fungus garden for food, the identity of the monoculture has been controversial and a number of names have been suggested (Kreisel, 1972; Hervey *et al.*, 1977; Kermarrec *et al.*, 1986). However, the fungus *L. gongylophorus* (A. Möller) Singer described by Möller (1893) has recently been confirmed as the same by Fisher *et al.* (1994b), although there may be species variations (Muchovej *et al.*,1991). The leaf material imported into the ants' nests by the ant hosts is variable and depends on the plant species available and suitable for the ants in the vicinity of the nest. It is normally cut from living leaves that are still on the host plant. The use of flowers and fruits has also been observed and they have formed useful additions in laboratory nests.

It is well established that plants growing in the wild are covered by an extensive epiphytic fungal community (Dickinson, 1986), and wild plants also contain fungal endophytes (Petrini, 1986). A large variety of fungal species is therefore imported into the nest on a daily basis. In a healthy nest, growth of these species is suppressed by a chemical cocktail which includes myrmicacin and phenylacetic acid. Möller (1893) appears to have been the first person to describe fungi other than *L. gongylophorus* in ant fungus gardens. He mistakenly associated a conidial state of an unnamed hyphomycete with the ant fungus *L. gongylophorus* but nevertheless accurately drew the conidial state of *Escovopsis weberi* Muchovej, Della and Lucia (syn. *Phialocladus zsoltii* Kreisel), a contaminant reported from the fungus gardens of *A. cephalotes* (Möller, 1893; Carmichael *et al.*, 1980; Powell, 1984). Although numerous workers have periodically reported filamentous fungi and yeasts in attine ant nests (Goetsch and Stoppel, 1940; Stahel and Geijskes, 1941; Weber, 1957; Craven *et al.*, 1970; Sihanonth *et al.*, 1973; Kermarrec *et al.*, 1986), there appears to have been no systematic study to evaluate the species and numbers that may be involved.

The research described here was designed to record filamentous fungi and yeasts in three nests of *A. cephalotes* and to examine whether the fungi in the nests corresponded with the epiphytic and endophytic mycota of the food plants. Further, it was investigated whether there might be fungi other than *L. gongylophorus* that had adapted to the environment of the nests and taken up permanent residence there, and whether environmental pressure on the colony such as an enforced diet might have adversely affected the ability of the ants to maintain a monoculture.

The fungus gardens of three colonies of *A. cephalotes* were sampled. Two of these (Q1 and Q2), collected from the field as 1-year-old colonies in the Arima

Valley, Trinidad, West Indies in 1991, were subsequently maintained in the laboratory at Exeter at 25°C, in a 12:12 h light regime and relative humidity > 70%. At the time of sampling, both colonies comprised two fungus gardens of approximately 2 l each. Prior to sampling they had been subjected to a strict 3 months regime of the same *Quercus ilex* L. (holm oak) leaf material collected on the Exeter University campus to test whether the mycota of the leaves could be detected in the nests. Colony Q1 was subsequently supplied with leaves of *Rosa* sp. (hybrid tea) for a further 3 months (Q1R) and resampled for similar tests. The third colony (FT), which was much larger than the other two, was sampled in its natural environment, located in steeply sloping terrain among clumps of bamboo at Noel Trace, St Augustine, Trinidad. The principal foraging trail for the ants, at the time of sampling, extended for 100–200 m to a cultivated garden where leaf pieces were cut from several trees including *Persea americana* Mill. and *Codiaeum variegatum* (L.) Blume. Samples of fungus garden were obtained from a nest chamber exposed by digging, and of waste material by taking it from ants emerging from a number of nest entrances at the base of the nest. The fungus gardens of all the nests were sampled by dividing the nest approximately into an upper and lower region. Then 30 pieces of fungus garden (approximately 0.5 cm^3) consisting of many individual leaf units were sampled from each region and placed immediately into individual sterile glass tubes. Each of these 30 pieces was subsequently dissected into the separate leaf units which rarely measured more than 2 mm^3. These represented fragments of leaf material cut by the ants during the preparation of the fungus garden (Fig. 9.8). Möller (1893) also recorded this fine division of the substratum by the ants. Units of this divided material were taken at random and plated on 1.5% malt extract agar medium (Fig. 9.8) supplemented with 250 mg l^{-1} Terramycin (MEA + T) to inhibit bacterial growth. One hundred pieces each from the upper and lower nest regions were plated. In addition, 100 units of discarded waste from each nest were treated similarly. The process was repeated for nest FT, except that 4 days elapsed between collection and leaf particle plating. Leaf units were incubated for up to 6 weeks at 20 ± 2°C depending on the growth rates of the fungi that emerged (Fig. 9.9). Fungi other than the ant food fungus, *L. gongylophorus*, were isolated by transferring hyphal tips to 2% MEA plates. Near-UV light (Philips TL 40W/05) was used for up to 3 weeks to induce sporulation.

Identification was normally possible after sporulation. *L. gongylophorus*, which has no known anamorph, could be readily identified on the original culture plates by its typical white colonies and bunches of gongylidia (Fig. 9.9). Leaves of *Q. ilex* fed to colonies Q1 and Q2 only, leaves of *Rosa* sp. fed to colony Q1 only, and leaves of *P. americana* and *C. variegatum* collected from the trees being used by the ants as food plants in Trinidad were screened separately for a qualitative presence of epiphytic and endophytic fungi. For the former, three leaves of each species were agitated in 100 ml of sterile water for 1 h, and 0.1 ml aliquots were pipetted on to three Petri dishes containing solid medium 1.5% MEA + T. For the endophytes, leaves were surface sterilized using sodium

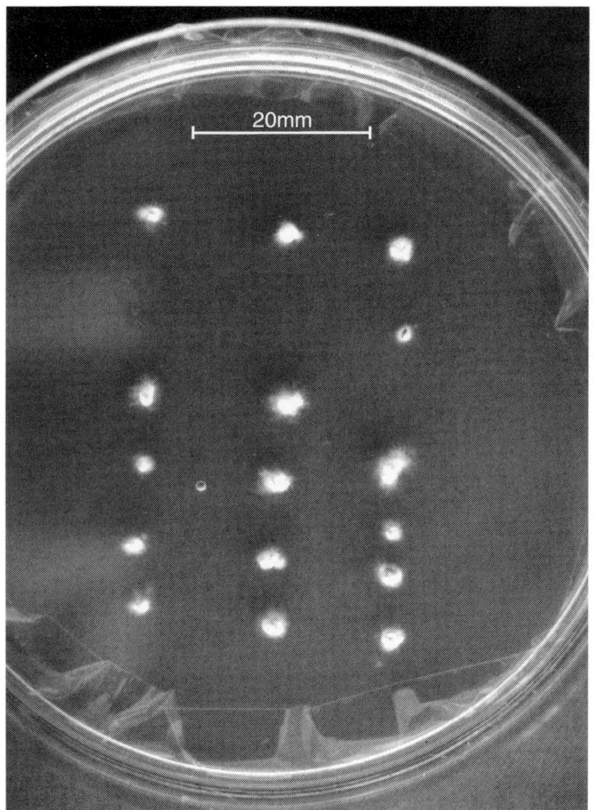

Fig. 9.8. Leaf fragments cut by the ants plated on nutrient agar growing only *L. gongylophorus* after 4 weeks. Scale bar = 20 mm.

hypochlorite and ethanol as described by Fisher *et al.* (1994a) . An estimate of the fungi that might be carried into the nest on the bodies of the ants was made by allowing ten of the leaf-bearing foragers to traverse ten separate 1.5% MEA + T plates for 2 h and then incubating the plates at 20°C for 5 days. The relative abundance of filamentous microfungi in the upper and lower regions of the fungus gardens from the three colonies sampled were compared using χ^2 analysis.

When considering the results of this survey, it is important to recognize that the attine fungus garden provides the principal larval food source in the form of gongylidia, and more rarely basidiome tissue of *L. gongylophorus*. The attine nests do not appear to be axenic cultures of *L. gongylophorus* (Table 9.1) if it is assumed that at least some of the fungi isolated from the nests were present in the nests in a mycelial form and not only as ungerminated conidia. When individual leaf fragments that made up the fungus garden were plated on to MEA + T (Fig. 9.8), a substantial proportion (21–53%) gave rise to cultures of only *L. gongylophorus* (Table 9.1). Many of these came from the upper section of the fun-

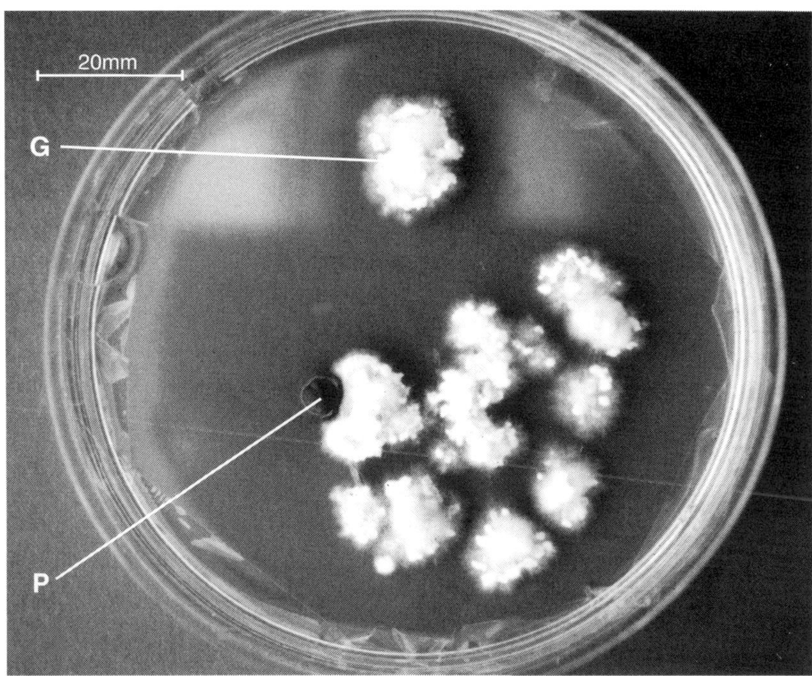

Fig. 9.9. Eight-week-old cultures of *L. gongylophorus* growing from leaf fragments showing clumps of hyaline gongylidia formations (arrowed G). The arrowed colony (P) is *Phomopsis quercella,* an endophyte from the leaves of *Q. ilex*. Scale bar = 20 mm.

gus gardens (Table 9.1), which was remarkable considering that these may have been placed in position by the ants only 24 h previously.

Möller (1893) also commented on rapid colonization of leaf fragments by the basidiomycete in field colonies. The highest colonization frequency of *L. gongylophorus* was recorded from nest Q1R (94%), the lowest from the field nest FT (72%). The latter may be partly due to the 4 day delay between sampling in Trinidad and particle isolation in the laboratory at Exeter. This would have led to some deterioration of the *L. gongyloporus* culture because of the overgrowth of other fungi in the absence of ant husbandry. That isolated pieces of fungus garden are subject to rapid deterioration was recorded previously by Fisher *et al.* (1994b), who failed to grow basidiomes of *L. gongylophorus* to maturity after they had been removed from a laboratory nest with large portions of the fungus garden because of the rapid overgrowth of fungal contaminants. This, and experimental evidence that *L. gongylophorus* shows no signs of antagonism against other fungi in laboratory culture (Stradling and Powell, 1986), suggests that, unlike many basidiomycetes, *L. gongylophorus* has a low capacity for allorecognition and is probably entirely dependent on ant husbandry to suppress incoming microfungi. Möller (1893) made similar observations on isolated fungus garden fragments and in abandoned nests.

Table 9.1. Frequency of occurrence of fungal species on 100 leaf fragments from the upper (U) and lower (L) regions of fungus gardens from three colonies of *Atta cephalotes* (FT, field collected (Trinidad); Q1, Q2, laboratory colonies supplied with *Q. ilex* leaves; Q1R, colony Q1 supplied for 3 months with *Q. ilex* and subsequently for 3 months with leaves of *Rosa*).

	FT		Q1		Q1R		Q2		
	U	L	U	L	U	L	U	L	W
Leucoagaricus gongylophorus	63	74	92	68	91	97	87	81	
(A. Möller) Singer	(37)	(42)	(42)	(0)	(56)	(51)	(60)	(19)	
Trichoderma hamatum (Bonorden) Bainier	4	2	0	0	0	0	0	0	+
Cladosporium herbarum (Pers.) Link	0	2	0	1	0	0	0	0	+
Cladosporium oxysporum Berk. & Curtis	1	0	0	1	2	0	0	0	+
Phoma nebulosa (Pers. & Fr.) Berk.	0	2	0	0	0	0	0	0	+
Phyllosticta ghaesembillae Koord.	17	6	0	0	0	0	0	0	
Glomerella cingulata: (Stonem.) Spaulding & V. Schrenk	3	4	0	0	0	0	0	0	
Trichoderma longibrachiatum Rifai	2	0	0	0	0	0	0	0	
Aureobasidium pullulans (de Bary) Arnaud	0	0	1	4	0	1	0	3	+
Fusarium solani Schlecht.	0	0	0	3	0	0	0	0	+
Aspergillus niger v. Tieghem	0	0	0	2	0	0	0	0	+
Epicoccum nigrum Link	0	0	2	4	0	0	2	6	+
Gliomastix murorum (Corda) Hughes	0	0	4	33	0	0	2	16	
Phomopsis ilicina v.d. Aa ined. syn. *Phyllosticta ilicina* Sacc.	0	0	0	3	0	0	0	4	
Phomopsis quercella (Sacc. & Roum.) Died.	0	0	0	4	0	0	3	0	
Phomopsis glandicola (Lév.) Gonz. Frag.	0	0	0	0	0	0	2	0	
Fusarium sp.	0	0	0	0	0	2	0	0	+

Phomopsis culmorum (W.G.Sm.) Sacc	0	0	0	0	0	0	0	6	+
Yeasts	19	33	38	41	35	42	24	27	+
Sterile mycelia	15	9	10	2	2	4	9	19	+
Sterile leaf fragments	2	0	3	2	5	0	2	1	
No. of species per leaf fragment	1.32	1.32	1.50	1.72	1.35	1.46	1.31	1.63	
No.of species per leaf fragment less yeasts	1.12	0.93	1.12	1.27	0.95	1.04	1.05	1.35	

W, species found in both nest and waste (+).
Figures in parenthesis are leaf fragments occupied by *L. gongylophorus* only.
Fungi found in waste only: Field fungus garden collection (Trinidad): *Aureobasidium pullulans, Cylindrocarpon destructans* (Wollenw.) C. Booth, *Fusarium pallidoroseum* (Cooke) Sacc.; *Phialocladus zsoltii* Kreisel; *Phomopsis persea* Zerova. Fungus garden Q1: *Penicillium* sp., *Rhizopus stolonifer* (Ehrenb.) Lind, *Trichoderma harzianum* Rifai. Fungus garden Q1R: *Aspergillus flavus* Link, *Cladosporium herbarum, Cladosporium sphaerospermum* Penz., *Syncephalastrum racemosum* Cohn : J. Schröt. Fungus garden Q2: *Trichoderma harzianum, Epicoccum nigrum.*

The ant–fungus association is an obligate mutualism and the maintenance of the fungus culture is essential to the survival of the colony. The leaf material, which largely makes up the plant portion of the nests of attine ants, is nutritionally valuable to endophytic and epiphytic fungi. This often variable material serves as the substratum for *L. gongylophorus*, and also renders the nest attractive to a wide range of microfungi, which are introduced with the leaves and on the bodies of the ants as spores and mycelium. Biologically active secretions by the workers' metapleural glands have been shown to suppress the germination of fungal spores (Schildknecht and Koob, 1971) and are one element of nest 'hygiene'. Another is the removal of refuse by workers which collect particles of refuse in the infrabuccal pocket and eject them outside the nest (Bass, 1993).

The identity of microfungi from outside the nest reflects mostly the epiphytic and endophytic fungal community of the plants collected by the ants as a fungal resource. This is suggested by the similarities between the taxa isolated from the colonies of Q1 and Q2, both supplied with only *Q. ilex* leaves, and the differences in taxa composition in Q1R and FT which had different leaf supplies (Tables 9.1 and 9.2). Seventeen taxa, a number of sterile mycelia, and yeasts were isolated from the fungus gardens (Table 9.1). There is no evidence to suggest that any of the filamentous fungi have adapted to become permanent colonizers of the fungus gardens. For example, *Gliomastix murorum* (Corda) Hughes, recorded as an epiphyte from leaf washings of *Q. ilex*, was isolated 37 times from the fungus garden Q1 but was absent in the same nest Q1R after the substratum had been changed to *Rosa* sp. *Cladosporium herbarum* (Pers.) Link and *Cladosporium oxysporum* Berk. & Curtis isolated from nests FT and Q1R (Table 9.1) are known to be cosmopolitan in distribution and occurred as endophytes or epiphytes in the leaves of most of the resource plants (Table 9.2). Both *Phomopsis ilicina* (Aa ined., syn. *Phyllosticta ilicina* Sacc.) and *Phomopsis quercella* (Sacc. & Roum.) Died., present in nests Q1 and Q2 but not Q1R, were also recorded as endophytes in *Q. ilex* leaves (Table 9.2). They have previously been recorded as among the four most frequently occurring endophytes in leaves of *Q. ilex* (Fisher *et al.*, 1994a). These two fungi are usually specific to *Q. ilex* leaves so they were probably imported into the fungus garden with the leaves. Other fungi recorded as epiphytes and endophytes of the food plants and also present in the fungus garden are listed in Table 9.2. Only *E. weberi* (syn. *Phialocladus zsoltii*; Muchovej and Della Lucia, 1990), reported repeatedly from the waste debris of *Atta* nests in the wild (Möller, 1893; Carmichael *et al.*, 1980; Powell, 1984), has also been recorded from waste in this study (Table 9.1). Recently, another species of *Escovopsis*, *E. aspergilloides* Seifert, Samson & Chapela, has been recorded from nests of *Trachymyrmex ruthae* Weber in Trinidad (Seifert *et al.*, 1995).

A comparison between the mycota in nests Q1 and Q1R showed qualitative and quantitative changes in species composition when the substratum of the nest was changed from *Q. ilex* to *Rosa* sp. The former is due to the importation of different fungal species from the rose leaves (Table 9.1). The change in substratum was also accompanied by a significant reduction in the frequency of fil-

Table 9.2. Some of the fungi isolated as endophytes and epiphytes from fresh leaves, used by the ants to supply the fungus gardens, and subsequently isolated from the fungus gardens.

Leaf species	Endophytes	Epiphytes
Quercus ilex	*Phomopsis ilicina*	*Cladosporium herbarum*
	Phomopsis quercella	*Aureobasidium pullulans*
	Aureobasidium pullulans	*Gliomastix murorum*
	Epicoccum nigrum	Yeasts
	Yeasts	
Persea americana	*Phyllosticta ghaesembillae*	*Fusarium solani*
	Phoma nebulosa	*Cladosporium oxysporum*
	Glomerella singulata	*Aureobasidium pullulans*
	Yeasts	Yeasts
Codiaeum variegatum	*Phyllosticta ghaesembillae*	*Cladosporium herbarum*
	Aureobasidium pullulans	*Aureobasidium pullulans*
	Yeasts	Yeasts
Rosa (hybrid tea)	*Aureobasidium pullulans*	*Cladosporium sphaerospermum*
	Cladosporium herbarum	

amentous fungi in both upper and lower regions of the fungus garden (χ^2 = 8.91, $P < 0.01$; $\chi^2 = 54.07$, $P < 0.001$, respectively). This reduction in the total numbers may be due to a smaller number of fungi imported into Q1R with the rose leaves, but it could also reflect the improved 'colony fitness' after the diet change from the dry and tough foliage of *Q. ilex* to the more succulent leaves of *Rosa* sp. 'Colony fitness' is defined here as the productivity of the colony in terms of importation of leaf material and fungus garden construction. In laboratory colonies this switch between food plants was characterized by a marked improvement of 'colony fitness' (D. Stradling, Exeter, 1990, personal communication). In the field, *A. cephalotes* colonies collect material from a wide range of tropical rainforest trees (Cherrett, 1968) and are characterized by frequent switching between plant species, but nothing is known as to how this reflects on 'colony fitness'. Comparisons of the incidence of filamentous microfungal contaminants between the upper and lower portions of each nest and between colonies revealed significant heterogeneity. For example, higher numbers were recorded from the lower portions of nests Q1 and Q2 but the reverse for colony FT (Table 9.1). There are many unknown variables that could account for this anomaly. One possibility is that in the cleaner environment of the laboratory fewer contaminants are present on the material imported to the upper portion of the fungus garden where the ants always first deposit new plant material. Furthermore, leaves from temperate vegetation contain fewer endophytes than those from the tropics (Fisher *et al.*, 1993).

The presence of yeasts was widespread in all the fungus gardens (Table 9.1), an observation in agreement with that made previously by electron microscopy of fungus gardens of *A. cephalotes* by Craven *et al.* (1970). For the four ant colonies sampled, between 20 and 40% of leaf fragments showed the presence of yeasts, but no attempt was made to distinguish different taxa. Craven *et al.* (1970), using SEM techniques, found that imported leaf material lacked detectable yeast cells. However, the widespread occurrence of yeasts on leaves collected from the field (Table 9.2), and their recovery from nests (Table 9.1) suggests that they are imported into the nest on the leaf substratum.

In separate experiments, when ants were allowed to traverse agar plates for 2 hours, both filamentous fungi and yeast contaminants were evident. *Cladosporium* spp. were present on nine out of ten plates (3.10 ± 0.48 colonies per plate), *Aureobasidium* spp. on six out of ten plates (1.10 ± 0.35 colonies per plate) and numerous yeast colonies on all ten plates. This suggests that most other fungi were not imported into the nests on the ants but on the leaves. It would be reasonable to suppose that fungi such as *Cladosporium* spp. and *Aureobasium* spp., both of which readily produce large numbers of conidia, are brought into the nest as spores. Endophytes such as *P. ilicina* and *P. quercella* may survive the physical and chemical cleaning processes by the ants for varying periods as mycelium within the leaves.

Yeast trails were frequent on the test plates, suggesting that these fungi are frequently carried on the ants and therefore have the potential for widespread dispersal within the fungus garden matrix. This has been separately confirmed by the leaf piece assays (Table 9.1) and by Craven *et al.* (1970).

The use of inert markers, such as spines of the leaves of *Ilex aquifolium* which the ants deposit in the nests, has demonstrated that the ant fungus garden is a dynamic entity where new nest material is added at the top and over a period of 5–7 weeks descends to the base to be discarded eventually by the ants as waste. Thus there is a complete turnover of the fungus garden material within approximately 7 weeks, and the ultimate fate of all the microfungi attached to the substratum is as discarded waste (Table 9.1). This offers an explanation for the earlier mentioned disappearance of *G. murorum*, isolated 37 times from the fungus garden Q1 but lost from the same fungus garden (Q1R) after the substratum was changed.

Of the 300 waste fragments collected from just outside the three original fungus gardens (FT, Q1, Q2), none grew *L. gongylophorus*. It appeared to have been replaced by numerous fungi which for the most part differed from those in the gardens (Table 9.1), a finding in accord with Craven *et al.* (1970). In comparison, leaf fragments of the fungus garden that were actively tended by the ants yielded fewer filamentous contaminants per fragment. However, the total disappearance of *Leucoagaricus* from the waste fragments once again underlines the poor competitive ability of this basidiomycete.

The mycelial dynamics of *L. gongylophorus* in relation to the fungal contaminants which may enter the nest are unknown. It would be desirable to study them, although the difficulties of doing so would be considerable, partly because each contaminant is likely to interact differently with *L. gongylophorus* and also because any disturbance of the nest would affect the ant husbandry.

Acknowledgements

We thank Professor D.N. Pegler for confirming the identification of *L. gongylophorus* and Dr C.R. McDavid for collection and identification of leaf material in Trinidad. The original published source for Figures 9.1–9.8 was *Mycological Research* 98, pp. 884–888, 1994 and 100, pp. 541–546, 1996.

References

Bass, M. (1993) Studies on the ant–fungus mutualism in leafcutting ants (*Formicidae, Attini*). PhD thesis, University of Wales, Bangor, UK.

Belt, T. (1874) *The Naturalist in Nicaragua*. John Murray, London, pp. 63–65.

Carmichael, J.W., Bryce Kendrick, W., Conners, I.L. and Sigler, L. (1980) Form genera. In: *Genera of Hyphomycetes*. The University of Alberta Press, Edmonton, Alberta, Canada, p. 147.

Cherrett, J.M. (1968) The foraging behaviour of *Atta cephalotes* L. (Hymenoptera, Formicidae). 1. Foraging pattern and plant species attacked in tropical rain forest. *Journal of Animal Ecology* 37, 387–403.

Craven, S.E., Dix, M.W. and Michaels, G.E. (1970) Attine fungus gardens contain yeasts. *Science* 169, 184–186.

Dickinson, C.H. (1986) Adaptations of microorganisms to climatic conditions affecting aerial plant surfaces. In: Fokkema, N.J. and van den Heuvel, J. (eds) *Microbiology of the Phyllosphere*. Cambridge University Press, Cambridge, pp. 77–100.

Fisher, P.J., Petrini, O. and Sutton, B.C. (1993) A comparative study of fungal endophytes in leaves, xylem and bark of *Eucalyptus* in Australia and England. *Sydowia* 45, 338–345.

Fisher, P.J., Petrini, O., Petrini, L.E. and Sutton, B.C. (1994a) Fungal endophytes from the leaves and twigs of *Quercus ilex* L. from England, Majorca and Switzerland. *New Phytologist* 127, 133–137.

Fisher, P.J., Stradling, D.J. and Pegler, D.N. (1994b) *Leucoagaricus* basidiomata from a live nest of the leaf-cutting ant *Atta cephalotes*. *Mycological Research* 98, 884–888.

Goetsch, W. and Stoppel, R. (1940) Die Pilze der Blattschneider Ameisen. *Biologisches Zentralblatt* 60, 393–398.

Heim, R. (1957) A propos du *Rozites gongylophora*. *Revue de Mycologie* 22, 293–299.

Hervey, A., Rogerson, C.T. and Leong, I. (1977) Studies in fungi cultivated by ants. *Brittonia* 29, 226–236.

Kermarrec, A., Decharme, M. and Febvay, B. (1986) Leaf-cutting ant symbiotic fungi: a synthesis of recent research. In: Lofgren, C.S. and Van der Meer, R.K. (eds) *Fire Ants and Leaf-cutting Ants, Biology and Management*. Westview Press, Boulder, Colorado, pp. 231–246.

Kreisel, H. (1972) Pilz aus Pilzgarten von *Atta insularis* in Kuba. *Zeitschrift für Algemeine Mikrobiologie* 12, 643–654.

Maschwitz, U., Koob, K. and Schildknecht, H. (1970) Ein Beitrag zur Funktion der Metathoracaldrüse der Ameisen. *Journal Insect Physiology* 16, 387–404.

Möller, A. (1893) Die Pilzgärten einiger südamerikanischer Ameisen. In: Schimper, A.F.W. (ed.) *Botanische Mittheilung Tropen* Heft VI. Verlag von Gustaf Fischer, Jena, Germany, pp. 1–127.

Muchovej, J.J. and Della Lucia, T.M.C. (1990) *Escovopsis*, a new genus from leaf-cutting ant nests to replace *Phialocladus* nomen invalidum. *Mycotaxon* 37, 191–195.
Muchovej, J.J., Della Lucia, T.M.C. and Muchovej, R.M.C. (1991) *Leucoagaricus weberi* sp.nov. from a live nest of leaf-cutting ants. *Mycological Research* 95, 1308–1311.
Petrini, O. (1986) Taxonomy of endophytic fungi of aerial plant tissues. In: Fokkema, N.J. and van den Heuvel, J. (eds) *Microbiology of the Phyllosphere*. Cambridge University Press, Cambridge, pp. 175–187.
Powell, R.J. (1984) The influence of substrate quality on fungus cultivation by some attine ants. PhD thesis, University of Exeter, UK.
Schildknecht, H. and Koob, K. (1971) Myrmicacin, the first insect herbicide. In: *Angewante Chemie*, Vol. 10. Riande, Überlingen, pp. 124–125.
Seifert, K.A., Samson, R.A. and Chapela, I.H. (1995) *Escovopsis aspergilloides*, a rediscovered hyphomycete from leaf-cutting ants. *Mycologia* 87, 407–413.
Sihanonth, P., Michaels, G.E. and Dix, M.W. (1973) Selective inhibition of fungi in the attine fungus garden. Abstracts Anniversary Meeting American Society of Microbiology. *Journal of the American Society of Microbiology* 73, 139.
Singer, R. (1975) *The Agaricales in Modern Taxonomy*, 3rd edn. J. Cramer, Vaduz, Liechtenstein.
Singer, R. (1986) *The Agaricales in Modern Taxonomy*, 4th edn. Scientific Books, London.
Stahel, G. and Geijskes, D.C. (1941) Weitere Untersuchungen über Nestbau und Gartenpilze von *Atta cephalotes* L. und *Atta sexdens* L. *Entomological Review* 12, 243–268.
Stradling, D.J. and Powell, R.J. (1986) The cloning of more highly productive fungal strains: a factor in the speciation of fungus growing ants. *Experientia* 42, 962–964.
Weber, N.A. (1957) Weeding as a factor in fungus culture by ants. *Anatomical Record* 28, 638.
Weber, N.A. (1966) Fungus growing ants. *Science* 153, 587–604.

Conservation of Mycodiversity in India: an Appraisal 10

T.N. Kaul

249 Nilgiri Apartments, Alaknanda, New Delhi 110 019, India

Introduction

Conservation of natural resources, both biotic and abiotic, was appreciated from the early dawn of civilization 5000–7000 years ago; it was even an essential part of some major religions, especially Buddhism and Jainism over two millennia ago. Tivy (1993) recorded that conservation of forest resources was official policy in China as early as the 3rd century BC. Indian civilization also has a long history of conservation ethics starting from the Vedic period. Conservation of biodiversity is ingrained in Indian history, culture, religion and philosophy. Existence of sacred forests in some parts of India is evidence of the Indian people's deep interest in conservation.

Conservation of biological diversity has been the subject of intense debate all over the world during the past two decades or so and is considered critical to the health and stability of the biosphere. It has been defined by the International Union for Conservation as human use of the biosphere so that it may yield the greatest sustainable benefit to the present generation while maintaining its potential to meet the needs and aspirations of the future.

Terrestrial biodiversity exists on earth in eight broad realms and 193 biogeographical provinces (Khoshoo, 1991). The distribution, however, is uneven; tropical and subtropical regions (mostly developing countries) are vastly richer than the temperate and polar regions (developed countries; Hawksworth, 1991); although tropical forests cover only 7% of the earth's land, they contain 50% of all species (McNeely *et al.*, 1990).

The Indian Scene

The Indian subcontinent is a vast land mass covering 3,287,263 km^2 (Oxford School Atlas, 1993), lying between latitudes 8–37°N and longitudes 68–98°E (Fig. 10.1). The country has a coastline over 7516 km long (Khoshoo, 1996a). The altitude varies from sea level to 7817 m, and it is isolated from the rest of Asia by the mountain ranges of Himalaya in the north. The subcontinent falls into two out of eight proposed global realms, the Palaearctic and Indo Malayan realm and 12 biogeographical provinces according to one classification

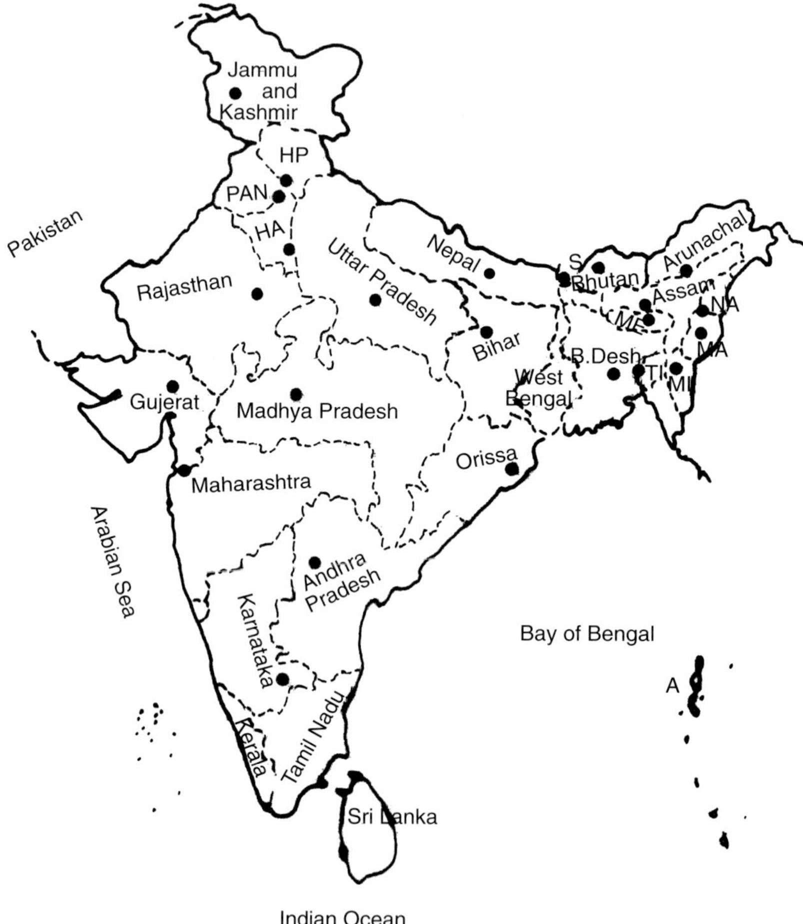

Fig. 10.1. Map showing location of various Indian states, union territories and adjacent countries. Abbreviations: A, Andaman and Nicobar islands; B, Desh, Bangla Desh; HA, Haryana; HP, Himachal Pradesh; MA, Manipur; ME, Meghalaya; MI, Mizoram; NA, Nagaland; PAN, Punjab; S, Sikkim; TI, Tripura.

(Khoshoo, 1996a). Khoshoo (1991) further identified six broad terrestrial ecosystem types in these provinces: tropical humid forest; tropical dry or deciduous forest (including monsoon forest or woodlands); warm deserts and semi deserts; cold winter (continental) deserts; mixed mountain and highland systems (with complex zonation); and mixed island systems. However, Rodgers and Panwar (1988), revising the earlier classifications, divided the Indian subcontinent into 10 biogeographical zones and 25 biotic provinces. These zones are: trans-Himalayan (Ladakh), Himalayan, desert, semi-desert, Western Ghats, Deccan plateau, gangetic plains, north-east India, islands and coasts (Khoshoo, 1996a). Champion and Seth (see World Wide Fund for Nature – India, 1992) recognized 16 major forest types with various subdivisions in India covering 63.8 Mha (19.4% of geographical area). Each type differs significantly from the others in structure and floristics.

The total number of living species recorded for India is presented in Table 10.1. Nearly 73% of India's bio-wealth consists of fungi (19.5%; including lichenized taxa), insects (40%) and angiosperms (13.5%). This tallies generally with the overall trend seen in the tropics and subtropics (Khoshoo, 1995). Although India has only 2.4% of the land area of the world as a whole, according to the present estimates, it hosts 8% of the global biodiversity.

Table 10.1. Biota of India (after Khoshoo, 1995).

	Number of species	Percentage
Bacteria	850	0.7
Algae	2,500	2.0
Fungi (non-lichenized taxa)	23,000[a]	18.2
Fungi (lichenized taxa)	1,600	1.3
Bryophyta	2,700	2.1
Pteridophyta	1,022	0.8
Gymnosperms	64	0.1
Angiosperms	17,000	13.5
Protozoa	2,577	2.0
Mollusca	5,042	4.0
Crustacea	2,970	2.4
Insecta	50,717	40.0
Other invertebrates including Hemichordata	11,252	9.0
Protochordata	116	0.1
Pisces	2,546	2.0
Amphibia	204	0.2
Reptilia	428	0.3
Aves	1,228	1.0
Mammalia	372	0.3
Total	126,188	100.0

[a]Varma and Sarbhoy (1996) estimated the number of known species in India to be *c.* 24,000.

Geographically four main regions have been defined in the subcontinent. These are: the Himalaya, the Indo-gangetic plain, the Indian desert and Deccan plateau. In view of the importance of vascular plants in any assessment of macrofungal diversity, the vegetational patterns and climatic diversity of these areas are summarized.

The Himalaya

The Himalaya represents the highest mountain system in the world extending across northern India from west to east, it is 2500 km long, occupying an area of 236,000 km^2. It extends from the Indian plains to the Tibetan Highlands in four almost parallel ranges from south to north. These are: the outer Himalaya or Siwalik range, the middle or lesser Himalaya, the inner or great Himalaya and the trans-Himalayan region.

The western and eastern flanks of the Himalaya differ. The western Himalayan ranges are much wider and colder with a drier climate. In contrast, the eastern ranges are among the wettest regions of the world with great biodiversity (Khoshoo, 1996b). The western ranges have a vegetation which is drought resistant. There are large populations of conifers such as the chir and blue pine, deodar, fir and spruce. On the other hand, in the east, conifers though present are not a dominant element. *Rhododendron* dominates the plant life of this region. The eastern ranges receive the full thrust of monsoon winds from the Bay of Bengal and contain a large number of epiphytes, and a profusion of orchids, tree ferns, oaks, magnolias and gymnosperms not found in western Himalaya: *Larix*, *Tsuga dumosa*, *Podocarpus* and *Gnetum*. Sixty-three species of bamboo are also found in this region, there being very few in western Himalaya.

Khoshoo (1992) describes the forest types found in both regions. In western Himalaya the vegetation can be distinguished into six different types: tropical deciduous forests, subtropical pine forests, Himalayan moist temperate forests, Himalayan dry temperate, subalpine forests, and alpine pastures and scrub.

In the eastern Himalaya, which is a region of high monsoon intensity with a humid climate, the vegetation types are: tropical evergreen forests, subtropical forests, eastern temperate forests, alpine vegetation, and stony deserts.

The Indo-gangetic plain

The Indo-gangetic plain covers 643,700 km^2 formed by the river basins of the Indus, Ganges and Brahmaputra (Manjula, 1983). This area is largely tropical, with a mean annual temperature of 26°C. Much of the region has a high rainfall. In Assam, a tropical rainforest vegetation exists, elements of which are shown to have affinities with the forests of Kerala in the far south. The Indo-gangetic plain is an area of high population and extensive cultivation, so little collecting of fleshy fungi has been undertaken apart from in Assam and the Calcutta region of West Bengal.

The Indian desert

The Indian desert is a region to the north-west of India covering Rajasthan and has minimal rainfall and few macrofungi.

The Deccan plateau

The Deccan plateau is the oldest land mass in India (Manjula, 1983). In the north it is bounded by the Vindhya mountains with an altitudinal range of 450–1200 m and is separated from the Arabian Sea Coast and Bay of Bengal Coast by the Western Ghats and Eastern Ghats respectively. The ghats have an altitudinal range of *c.* 500–650 m. The Western and Eastern Ghats meet at Nilgiri Hills in Tamil Nadu State, and in the areas of highest elevation a sub-temperate mycobiota is to be found. A major part of the Indian subcontinent is, however, governed by the tropical monsoon and the plains of southern India experience periods of high temperature and humidity with a mycobiota associated with that climate.

Conservation Initiatives

Despite compassion for all life, ingrained in the psyche of most Indians from childhood, considerable loss of biodiversity, especially wild animal life, has occurred in India. However, recently, a move to conserve biodiversity has started, although microbial biodiversity still receives low priority.

For any meaningful conservation programme the following steps are essential. First, an assessment of biodiversity has to be made, together with a study of changes over a specified period. Subsequently, lists of threatened or extinct species should be compiled (Red Data Lists). In the final phase, conservation measures, both *in situ* and *ex situ*, have to be drawn up and implemented.

The main agencies for assessment of biodiversity in India are the Botanical Survey of India (established in 1880) and the Zoological Survey of India (established in 1916). Other institutions involved in ecological surveys and research are: the Forest Research Institute, Dehra Dun; the Wild Life Institute of India, Dehra Dun; the Gobind Ballabh Pant Institute of Himalayan Environment and Development at Almora (UP), and the National Environment Engineering Research Institute at Nagpur. Besides the above, institutional support in this activity also comes from universities (both agricultural and traditional) and national institutes, including the Indian Agricultural Research Institute, New Delhi, and the Indian Institute of Science, Bangalore. The Ministry of Environment and Forests recently established by the Government of India can be considered to be a 'nodal' agency for conservation measures.

Among the non-governmental agencies, the World Wide Fund for Nature–India (WWF–India), established in 1969, is the country's largest conservation agency. Other non-governmental organizations involved in promoting nature conservation in India are: the Bombay Natural History Society; the Indian National Trust for Art and Cultural Heritage (INTACH); the Centre for Environmental Education (CEE); the Centre for Science and Environment (CSE); and Development Alternatives (World Wide Fund for Nature–India, 1992).

Both *in situ* and *ex situ* measures currently operate in India to conserve biodiversity. Among the *in situ* measures are the development of a network of protected areas, including biosphere reserves, national parks and sanctuaries. Since the Wild Life (Protection) Act 1972, their number increased to 66 national parks and 421 sanctuaries by 1990, covering around 4% of the country's geographical area (World Wide Fund for Nature-India, 1992). Seven biosphere reserves had also been notified by 1991 (Khoshoo, 1991). *Ex situ* conservation is achieved by field gene banks, seed and other banks, botanical gardens and zoological parks. The situation regarding biodiversity conservation in Himalaya is discussed in detail by Khoshoo (1996b).

Among 17,000 species of flowering plants in India, 6850 (40%) are endemic to India, with *c.* 50% in the Himalaya (see Khoshoo, 1966b). Endemic flora and fauna assume great importance in determining priorities for conservation.

Macromycetes

The overall conservation activities in India have been summarized and are followed here by a résumé of the mycobiotic situation.

Conservation studies on fungal diversity are a recent development on the world scene, having started in the 1980s, although recording of fungi was started as early as 1729 by Micheli. Attention has been mainly devoted to macrofungi, commonly known as mushrooms. Watling (1990) observed that the chances of producing a Red Data List for microfungi are practically nil.

Fungal diversity is often assumed to be like biodiversity in general in tending to be greater in the tropics and subtropics than at higher latitudes (Lodge, 1997). Presently, only 4.6% of the world's estimated fungi are known and most of these are from temperate regions of the world (Hawksworth, 1991). Coddington (see Sarbhoy, 1997) considered that more than 20–40% of the tropical fungi are undescribed, but tropical forest ecosystems are undergoing massive annual losses.

History

There are references to the use of mushrooms as food and medicine in India in the ancient medical treatise, Charaka Samhita (3000 ± 500 BC). However, their scientific study is of recent origin. It can be said to have started with the identification and description of *Podaxis pistillaris* (L.: Pers.) Morse by Linnaeus in the 18th century from a collection sent to him by Koening from Tamil Nadu State. A subsequent collection made by Sir J.D. Hooker, mostly from Assam, Darjeeling, Sikkim and Khasi hills, led to the publication of a series of papers by an English mycologist, the Revd M.J. Berkeley between 1850 and 1882 (see Natarajan, 1995).

Collection and scientific study of larger fungi from India have been divided into three phases. The first phase lasted from 1825 to 1899 and, in addition to Berkeley and Montagne, recorders during this phase included Fries, Léveillé, Currey, Cooke, Massee, Watt and Lloyd (Sathe, 1979; Natarajan, 1995).

The second phase (1900–1969) started with Paul Henning's significant contribution in 1900 and 1901, describing another 32 genera and 68 species (Natarajan, 1995). A significant feature of the second phase was the involvement, besides European and American workers, of a number of Indian workers in research on larger fungi (Sathe, 1979). Special mention should be made of the work on Indian fungi by E.J. Butler at Pusa (Bihar) in the post of Imperial Mycologist. His efforts led to the production of the first authoritative list, *Fungi of India*, in collaboration with G.R. Bisby (Butler and Bisby, 1931). This publication has been continuously updated until the latest edition by Sarbhoy *et al.* (1996). Among the Indian workers of this period were Professor S.R. Bose (Calcutta, West Bengal) and Professor K.S. Thind (Punjab University, Chandigarh). The third phase of the work, which is continuing, can be said to have started in the early 1970s with the development of an edible mushroom industry in India, providing much needed impetus.

Manjula (1983) critically examined all agaricoid and boletoid records from India and Nepal housed in the Kew Herbarium, UK, and listed 538 valid species attributed to 115 genera and 20 families. This provided a baseline for subsequent assessment of agaric resources of these countries. Watling and Abraham (1992) put the total of agarics in India at *c.* 650 species. However, Sarbhoy (1997) observed that, whereas 450 mushroom species (*Agaricales*) had been reported by 1977, a similar number has been added either to regional or Indian subcontinent lists of mushrooms during the last decade, taking the total to *c.* 900. Watling (1978) and Watling and Gregory (1980), comparing it with British agaric records, considered this to be too low a figure.

Identification and description of fungi have been carried out in the past in a number of Indian universities (both traditional and agricultural), and in research institutes mainly under the auspices of the Indian Council of Agricultural Research (ICAR) and the Council of Scientific and Industrial Research (CSIR). A national effort at assessment of fungal diversity and consequent conservation measures was made with the development of an Advanced Centre for Biosystematics (Microbes and Invertebrates, ACBS) at the Indian Agricultural Research Institute (IARI), New Delhi in June 1993 (Varma and Sarbhoy, 1996). This centre has collaborated with the Commonwealth Agricultural Bureaux International (CABI). The ACBS is also developing an Indian Biosystematics Network (IBNET) which will foster links with biosystematics centres within various universities, ICAR Institutes and CSIR Institutes. The country also plans to establish a national Bureau of Agricultural Microorganisms which is expected to help in conserving biodiversity.

Current position at various Indian centres

Eastern Himalayan region

This region includes Eastern Himalaya proper and north-eastern hilly areas. The former includes the northern tip of West Bengal, northern parts of Assam, the whole of Arunachal Pradesh and Sikkim, and the latter the hilly states of Nagaland, Meghalaya, Manipur, Mizoram and Tripura (Khoshoo, 1992). Myers (1988, 1990, cited in Khoshoo, 1996a) listed 18 areas globally requiring special conservation efforts and designated them as 'hot spots'. Among these, two are in India, Eastern Himalaya and Western Ghats.

The first step in conserving the macrofungal diversity of a region would be the monitoring of mushroom species. Some efforts in this direction were made by groups working at Punjab University, Chandigarh, under the leadership of Professor K.S. Thind and Dr R.N. Verma and his associates at the Indian Council of Agriculture Complex in the north-eastern hill region at Shillong, established in the 1980s.

Sharma and Sidhu (1991) from Punjab University reported on the occurrence and distribution of the *Geoglossaceae* in the eastern Himalayan ranges of India. They maintained that the Himalaya in general and Eastern Himalaya and adjoining hills in particular are relatively rich in *Geoglossaceae*. They surveyed localities in and around West Bengal, Meghalaya, Assam and Arunachal Pradesh states and recorded 12 species distributed among nine genera: *Cudonia, Leotia, Maasoglossum, Microglossum, Mitrula, Thuemenidium, Spathularia, Trichoglossum* and *Geoglossum* with ecological notes. Two endemic species, *Maasoglossum verrucosporum* Thind & R. Sharma and *Mitrula agharkarii* Banerjee from the Himalayan range were also recorded by them. In India as a whole, the family is represented by 48 species within nine genera.

Thind and his associates have also worked on clavarioid homobasidiomycetes in the Himalaya, recording 181 taxa in 20 genera from Indo-Himalaya (Thind, 1961; Kaul, 1992). Sharda (1991) provided an updated check-list of these fungi with notes on their distribution. Genera reported from India included: *Amylaria, Clavariadelphus, Clavaria, Pterula* and *Ramaria*. In addition, Rattan and Khurana (1978) also published 'The Clavarias of Sikkim Himalaya'. Khoshoo, (1996b) concluded from Thind's researches that Himalaya was a major centre of diversity for higher fungi.

Verma *et al.* (1995) recorded the results of a macrofungal survey of the north-eastern hills (NEH), *c.* 8% of the country's total area, which can be divided into well-defined zones. These workers covered 102 locations in 24 districts of seven states including Sikkim and listed the species with ecological notes. Ninety-five species of larger fungi were determined and confirmed at the Kew Herbarium. Among these, 85 species were new records for the NEH region and 16 for India.

Lakhanpal (1993) comprehensively reviewed Himalayan *Agaricales*, listing all genera by family, and the number of species present in India and their distribution in both north-western and eastern Himalaya. Some genera were recorded only from Eastern Himalaya by one or more species, including *Cystoderma amianthinum* (Scop.: Fr.) Fayod, *Phaeogyroporus fragicolor* (Berk.) Horak, *Phylloporus pinguis* (Hook.) Singer, *Phylloporus sulphureus* (Berk.) Singer, *Tylopilus areolatus* (Berk.) Manjula, *Rhodocybe villosa* Horak and *Simocybe descedens* (Berk.) Manjula.

Mention should be made of the existence of sacred forests in Eastern Himalaya at Meghalaya (Khasi and Jainta Hills) and La Lygdoh at Mawsmai (Mawphlong and La Kyntok) near Cherapunji (Khoshoo, 1996b). These forests have been conserved over the ages on account of religious beliefs and have been left undisturbed, so diversity in these forests is unparalleled. Besides the angiosperm flora, rare species of macrofungi can also be expected. Bhagwat *et al.* (2000) presented relevant research in a poster session at the Liverpool Symposium on macrofungal diversity in three forested land use types including sacred groves in the Western Ghats of India. This showed that sacred groves contained the largest number of unique morphotypes whereas the forest reserve had none, suggesting their potential for conservation of macrofungal diversity in the Western Ghats of India.

Western Ghats (Korala and Maharashtra)

The Western Ghats, another 'hot spot' in India, is considered to be South Asia's last remaining tropical rainforest. Sathe (1980) described it as a flange of a high mountain range running along the western coast from Tapti Valley in the north to Cape Comorin in the south, with a total length of 1600 km and average elevation of 1000–1300 m with high peaks rising over 2400 m. It extends to three states of India, Maharashtra, Karnataka and Kerala.

Kerala's Western Ghats cover an area of approximately 20,000 km^2 and are the biologically richest tracts. The Western Ghats have an endemic higher plant flora of 1600 species (Khoshoo, 1994), and rank in importance equal to Eastern Himalaya for conservation initiatives.

According to Natarajan (1995), the entire region of southern India, comprising the four states of Tamil Nadu, Kerala, Karnataka and Andhra Pradesh, was neglected as regards studies on agarics until 1975. Natarajan (see Kaul, 1992) listed 115 species from Kerala, and a macrofungal survey of the area was carried out at the Plant Pathology Department of Kerala Agricultural University at Vellayani, Thiruvananthapuram, by Ms Bhavani Devi from 1985 to 1988. The collections were made from 12 agroclimatic zones in four monsoon seasons and revealed the presence of 134 species of mushrooms (including 14 gasteromycete species) belonging to 45 genera (Bhavani Devi, 1995). Analysing these findings, the macrofungi of Kerala were broadly grouped into pre-monsoon, monsoon and post-monsoon elements. The pre-monsoon fungi included species of *Agaricus*, *Coprinus*, *Lepiota*, *Leucocoprinus*, *Psathyrella*, *Mycena*, *Agrocybe* and *Boletus*. The monsoon element consisted of *Cortinarius*, *Schizophyllum*, *Marasmius* and *Pleurotus* whereas species of *Cortinarius*, *Pleurotus* and *Termitomyces* occurred post monsoon. Very few collections were made 30 days after the monsoon period. Edible fruit bodies included species of *Termitomyces*, *Volvariella*, *Pleurotus*, *Macrolepiota*, *Boletus* and *Calvatia*. *Tuber magnatum* Vitt. (*Ascomycotina*), a highly prized truffle, is regularly collected and consumed by tribal people in the forest area of the southern part of this state.

Staff at the Botany Department of Calicut University and the Tropical Botanic Garden and Research Institute at Thiruvananthapuram are also surveying macrofungi, but it is evident that assessment of the macromycetes of the region is at only a preliminary stage.

Mushroom recording in Maharashtra was neglected for a long time and only 21 species of agarics had been recorded from the state by 1967 (Kamat *et al.*, 1971). Intensive work in the region only began after 1974 when A.V. Sathe and his group, working at the Maharashtra Association for Cultivation of Sciences, published a series of papers mainly on *Agaricales* (Sathe and Rahalkar, 1978; Sathe, 1979; Sathe and Deshpande, 1979, 1980; Sathe and Kulkarni, 1987). A comprehensive list of 231 mushrooms recorded from all regions of Maharashtra state was published by Patil *et al.* (1995).

North-west Himalaya (Kashmir, Punjab and Chandigarh Union Territory and Himachal Pradesh)

Khoshoo (1991) identified 26 'hot spots' in India and maintained that the Himalayan belt as a whole constituted one 'mega-hot spot'. North-western Himalaya has been the centre of intensive research on larger fungi since the 1950s.

Berkeley in 1876 (see Watling and Gregory, 1980) was probably the first to report larger fungi from the Kashmir valley. Although fragmentary recording was made from time to time, a serious effort was initiated by the author and his group in the late 1960s, working at the Regional Research Laboratory, Srinagar,

Kashmir. Sustained work carried out by Kaul, Kachroo and Abraham for over two decades resulted in the recording of 262 larger fungi from the region, among which 226 taxa were agarics (Abraham, 1991). A significant contribution to this study was made by Professor Watling from Edinburgh, UK, who, besides providing constant guidance to these workers, published a list of 119 species of larger fungi from the area, based on his personal collection (Watling and Gregory, 1980).

There are two active centres of research on macrofungi in this region and collections have mostly been made from North-west Himalaya. The first centre, at Punjab University, Chandigarh, led by Professor K.S. Thind, produced a series of papers on operculate discomycetes, particularly *Pezizales*. A total of 226 operculate discomycetes have been recorded from India so far; the major contribution from Thind and his associates.

Kaushal (1991), from the same group, has made a systematic study of North-west Himalayan species of *Helvella* and recorded 11 species from this region. Detailed descriptive and mycoecological notes on eight species were included. Thind and his colleagues have also published a series of papers on the *Polyporaceae* of India, and Thind (1973) reviewed the status of aphyllophoroid taxa in India. Later Rattan (1977) recorded 198 species of resupinate aphyllophoroid taxa from North-west Himalaya and Rawla and Arya (1987) recorded five agarics from northern India.

Another centre involved is the Botany Department, Punjabi University, Patiala. Saini *et al.* (1988, 1989) and Atri *et al.* (1991a) have described many species of *Russula* and *Lactarius*. Atri *et al.* (1991b) contributed a paper on systematic studies of *Agaricus campestris* (L.) Fr., and Atri and Saini (1989) reviewed work on the *Russulaceae* worldwide including the Indian components. To date only 81 taxa (55 of *Russula* and 26 of *Lactarius*) have been recorded from India.

Three important centres of work on macrofungi in the state are: The Biosciences Department, University of Himachal Pradesh, Shimla; the Agricultural and Horticultural University, Solan, and the National Research Centre on Mushroom (ICAR), Solan. Professor Lakhanpal, working at the University of Himachal Pradesh, Shimla, has made a major contribution with a list of 190 species of *Agaricales* occurring over the entire North-west Himalayan region (Lakhanpal,1995). His list, however, was restricted to *Agaricales* only and he did not catalogue species recorded before 1982 that had been published by Manjula (1983).

South India

Natarajan started work at the Centre of Advanced Studies in Botany, University of Madras in 1975 and his group has been collecting from the entire southern and south-western region. They started a series entitled 'South Indian

Agaricales', publishing over two dozen papers. Natarajan (1995) presented a list of 230 agaric and bolete species distributed among 67 genera from southern Indian states excluding Kerala.

Indo-gangetic Plains (Punjab, Uttar Pradesh and Bengal)

These plains are distributed over a number of Indian states: Punjab, Haryana, Delhi, Uttar Pradesh, Madhya Pradesh, Bihar, Orissa, West Bengal and Assam, but there has been limited activity as regards the scientific study of macrofungi in this region. Mention, however, can be made of the contribution by some individuals or small groups.

Two groups, working at Punjabi University, Patiala and Punjab University, Chandigarh (referred to earlier), have been active in the study of macrofungi in the plains of Punjab. Saini and Atri (1995) listed 94 species in > 24 genera belonging to the *Agaricales* and gasteromycetes. Prominent genera were: *Morchella, Agaricus, Agrocybe, Conocybe, Coprinus, Bolbitius, Pleurotus, Lycoperdon* and *Calvatia*. They also stated that the overall picture as far as the mushrooms of Punjab were concerned was scant, and much effort would be needed to bring out the extent of its mushroom resources.

Another centre of work in the northern Indian plains has been at the National Botanical Research Institute, Lucknow. Pathak and Gupta (1979) reported 58 species of agarics from the Lucknow area distributed among 25 genera. Prominent genera were: *Agaricus, Amanita, Chlorophyllum, Coprinus, Macrolepiota, Pleurotus, Termitomyces* and *Volvariella*. Earlier, Pathak *et al.* (1978) recorded a total of 13 species of *Volvariella* from India.

Macrofungi in Bengal have been studied by Professor R.P. Purkayastha and Professor N. Samajpati, both working in the Botany Department, University of Calcutta. Purkayastha devoted attention to wild edible mushrooms of the region and succeeded in cultivating one of them, *Calocybe indica* Purkayastha & Chandra. Purkayastha and Chandra (1976, 1985) compiled lists of Indian edible mushrooms which included 283 species of larger fungi.

Rajasthan

Rajasthan is the second largest state of India and represents 10% of the land area of the country. However, about 57% of the state consists of the great Thar desert (Doshi and Sharma, 1997), which has been designated as one of the biosphere reserves and, as a representative of the desert ecosystem in India, is especially important from the conservation angle.

An intensive survey of wild mushrooms was conducted throughout the state by A. Doshi working at the Department of Plant Pathology, Rajasthan College of Agriculture for 8 years (1989–1996). Doshi and Sharma (1997) provided a detailed list of macrofungi occurring in the region with mycoecological notes. A total of 173 species belonging to 95 genera was recorded from the area.

Most genera (18) were gasteromycetes or aphyllophoroid taxa (17). Special mention should be made of two edible gasteromycetes, *Phellorinia inquinans* Berk. and *Podaxis pistillaris*, tonnes of which can be collected from desert areas. *P. inquinans* is associated with sand dunes in the area (Singh, 1994).

Discussion

The foregoing account has shown that the mycological resources of the country are far from fully investigated. Horak (see Lakhanpal, 1995) also pointed out that, considering ecological niches, the Himalayas are still mycologically little explored. Conservation measures, however, can be planned only after widespread recording, mapping and studies on changes in the mycobiota of the Indian subcontinent. The first step in this direction was taken when Dr M.S. Swaminathan (President of WWF–India) sanctioned the preparation of a project report: *Conservation of Mushroom Resources in India* (Kaul, 1992). This report, besides cataloguing the mushroom resources of the country, presented a plan of action for which Kaul (1993) provided the salient features. Unfortunately, no progress has been made to date. It is hoped that new initiatives taken with the commissioning of the Advanced Centre of Biosystematics (Microbes and Invertebrates) will bear fruit in years to come, as it will act as a nodal agency for the entire country.

Hawksworth (1992) emphasized the role of ectomycorrhiza as bioindicators and monitors of ecosystem health. Watling and Abraham (1992), reporting on ectomycorrhizal fungi of Kashmir, recorded that out of 175 species of agarics and clavarioid fungi in Kashmir, 77 were suspected of being mycorrhizal and were also found elsewhere in India. Lichens, too, have been found to be promising indicator species (Hawksworth and Rose, 1976), and it is essential that this group is not neglected in conservation studies. India is reported to contain 1600 species of lichens and Patwardhan (1983) provided a list of rare and endemic species of the Western Ghats.

Tribal involvement is also an important aspect of the conservation of mycological resources in India. Nearly 450 tribal communities constitute about 6% of the country's total population of > 60 million. They possess intimate knowledge of the mushroom resources of their area. Subramanian (1982) dwelt on the urgency of the study, as this information could be permanently lost with the extinction of these tribes.

Few attempts have been made to link the Indian mycota with those of other countries. Abraham (1991) in an overview of the Kashmir mycota maintained that some links with those of Europe and North America were apparent, and Natarajan (1995) considered that mycological recording in the last two decades revealed a lack of similarity between the agaric mycobiota of Tamil Nadu, Karnataka and Andhra Pradesh and Sri Lanka, although Sri Lanka is geographically similarly situated. It has more resemblance to that found in East

Africa and South America. It should be mentioned here that, as opposed to the Indian subcontinent, the agarics of Sri Lanka have been extensively studied since 1871, and as many as 433 species were enumerated as early as 1875 (see Natarajan, 1995).

Watling (1978) has discussed the subject in detail. He maintained that various elements expected in the Indian macromycota could be placed in four categories: (i) cosmopolitan species; (ii) possible South-eastern Asiatic elements; (iii) possible Mediterranean–European elements; and (iv) eastern Asiatic elements (Indo-Chinese).

The gamut of agaricoid species present in India has also been influenced by the replacement of natural forests by exotic trees; for example conifers, eucalypts and *Casuarina* in the hills of southern India. Introduced fungi, for example *Amanita muscaria* (L.: Fr.) J.D. Hook., have been brought in with such exotic tree species. These endemics cause identification problems due to the paucity of available mycologists in India and neighbouring countries.

Watling (1978), when outlining a fruitful future research programme for the study of agarics in India, made special mention of an analysis of a possible connection between the agaric communities of the Malagasy Republic and the records from Sri Lanka and southern India, and secondly of a survey of pockets of relict rainforest vegetation before they disappear. Cooperation with other countries would be necessary for investigations of this type of mycogeography.

Acknowledgements

The author is grateful to Professor Roy Watling, now retired from the Royal Botanic Garden, Edinburgh, UK, for critical scrutiny of this account. Thanks are also due to Dr T. N. Khoshoo, Tata Energy Research Institute, New Delhi, for useful discussions.

References

Abraham, S.P. (1991) Kashmir fungal flora – an overview. In: Nair, M.C. (ed.) *Indian Mushrooms 1991*. Kerala Agricultural University, Vellanikkara, India, pp. 13–24.

Atri, N.S. and Saini, S.S. (1989) Family *Russulaceae* Roze – a review. In: Trivedi, M.L., Gill, G.S. and Saini, S.S. (eds) *Plant Science Research in India*. Today and Tomorrow's Printers and Publishers, New Delhi, India, pp. 115–125.

Atri, N.S., Saini, S.S. and Mann, D.K. (1991a) Further studies on north-west Indian Agricaceae – Systematics of *Lactarius deliciosus* (Fr.) S.F. Gray. *Geobios New Report* 10, 106–111.

Atri, N.S., Saini, S.S. and Gupta, A.K. (1991b) Systematic studies on *Agaricus campestris* (L.) Fr. *Geobios New Report* 10, 32–37.

Bhagwat, S.A., Brown, N.D., Watkinson, S.C., Savill, P.S. and Jennings, S.B. (2000) Macrofungal diversity in three forested land use types: a case study from the western ghats of India. *Abstracts: Tropical Mycology 2000, Liverpool John Moores University, 25–29 April 2000.*

Bhavani Devi, S. (1995) Mushroom flora of Kerala. In: Chadha, K.L. and Sharma, S.R. (eds) *Advances in Horticulture* 13, *Mushroom*. Malhotra Publishing House, New Delhi, India, pp. 277–316.

Butler, E.J. and Bisby, G.R. (1931) *The Fungi of India*. Imperial Council of Agriculture Research India Scientific Monograph 1, Calcutta, India; revised Vasudeva, R.S. (1960).

Doshi, A. and Sharma, S.S. (1997) Wild mushrooms of Rajasthan. In: Rai, R.D., Dhar, B.L. and Verma, R.N. (eds) *Advances in Mushroom Biology and Production*. Mushroom Society of India, National Research Centre for Mushroom, Solan, Chambaghat (HP), India, pp. 105–127.

Hawksworth, D.L. (1991) The fungal dimension of biodiversity: magnitude, significance and conservation. *Mycological Research* 95, 641–655.

Hawksworth, D.L. (1992) Litmus test for ecosystem health: the potential of bioindicators in the monitoring of biodiversity. In: Swaminathan, M.S. and Jana, S. (eds) *Biodiversity – Implications for Global Food Security*. MacMillan India, Madras, pp. 184–204.

Hawksworth, D.L. and Rose, F. (1976) *Lichens as Pollution Monitors*. Edward Arnold, London.

Kamat, M.N., Patwardhan, P.G., Rao, V.G. and Sathe, A.V. (1971) *Fungi of Maharashtra*, Bulletin 1. Mahatma Phule Krishi Vidyapeeth, Rahuri (M.S.), India.

Kaul, T.N. (1992) *Conservation of Mushroom Resources in India: Introduction and Action Plan*. World Wide Fund for Nature – India, New Delhi.

Kaul, T.N. (1993) Conservation of Mushroom Resources in India. *Mushroom Research* 2, 11–18.

Kaushal, S.C. (1991) Systematics of N.W. Himalayan species of *Helvella* (operculate discomycete). In: Khullar, S.P. and Sharma, M.P. (eds) *Himalayan Botanical Researches*. Ashish Publishing House, New Delhi, India, pp. 61–75.

Khoshoo, T.N. (1991) Conservation of biodiversity in biosphere. In: Khoshoo, T.N. and Sharma, M. (eds) *Indian Geosphere Biosphere Programme: Some Aspects*. Har-Anand Publications, Vikas Publishing House Private, New Delhi, India, pp. 183–233.

Khoshoo, T.N. (1992) *Plant Diversity in the Himalaya: Conservation and Utilization*. Govind Ballabh Pant Institute of Himalayan Environment and Development, Kosi, Almora, India.

Khoshoo, T.N. (1994) India's biodiversity: tasks ahead. *Current Science* 67, 577–582.

Khoshoo, T.N. (1995) Census of India's biodiversity: tasks ahead. *Current Science* 69, 14–17.

Khoshoo, T.N. (1996a) India needs a national biodiversity conservation board. *Current Science* 71, 506–513.

Khoshoo, T.N. (1996b) Biodiversity in the Indian Himalayas: conservation and utilization. In: *Banking on Biodiversity – Report on the Regional Consultation on Biodiversity Assessment in the Hindukush Himalayas*. Organized by International Centre for Integrated Mountain Development (ICIMOD), 19–20 December 1995, Kathmandu, Nepal, pp. 181–255.

Lakhanpal, T.N. (1993) The Himalayan *Agaricales* – status of systematics. *Mushroom Research* 2, 1–10.

Lakhanpal, T.N. (1995) Mushroom flora of North-west Himalayas. In: Chadha, K.L. and Sharma, S.R. (eds) *Advances in Horticulture* 13: *Mushroom*. Malhotra Publishing House, New Delhi, India, pp. 351–373.

Lodge, D.J. (1997) Factors related to diversity of decomposer fungi in tropical forests. *Biodiversity and Conservation* 6, 681–688.

Manjula, B. (1983) A revised list of the agaricoid and boletoid basidiomycetes from India and Nepal. *Proceedings of the Indian Academy of Sciences (Plant Science)* 92, 81–213.

McNeely, J.A., Miller, K.R., Reid, W.V., Mittermeier, R.A. and Werner, T.B. (1990) *Conserving the World's Biological Diversity*. International Union for Conservation of Nature and Natural Resources, Gland, Switzerland.

Micheli, P.A. (1729) *Nova Plantarum Genera*. Florence, Italy.

Natarajan, K. (1995) Mushroom flora of South India (except Kerala). In: Chadha, K.L. and Sharma, S.R. (eds) *Advances in Horticulture* 13 *Mushroom*. Malhotra Publishing House, New Delhi, India, pp. 381–397.

The Oxford School Atlas (1993) 28th edn. Oxford University Press, Delhi, India.

Pathak, N.C. and Gupta, S. (1979) Agaric flora of Lucknow. In: Chattopadhyay, S.B. and Smajpati, N. (eds) *Advances in Mycology and Pathology*. Oxford and IBH Publishing Co., New Delhi, India, pp. 76–80.

Pathak, N.C., Ghosh, R. and Singh, M.S. (1978) The genus *Volvariella* Speg. in India. In: Kaul, T.N. and Atal, C.K. (eds) *Indian Mushroom Science* I. Indo-American Literature House, Globe, Arizona, and Canada, pp. 295–303.

Patil, B.D., Jadhav, S.W. and Sathe, A.V. (1995) Mushroom flora of Maharashtra. In: Chadha, K.L. and Sharma, S.R. (eds) *Advances in Horticulture* 13, *Mushroom*. Malhotra Publishing House, New Delhi, India, pp. 317–328.

Patwardhan, P.G. (1983) Rare and endemic lichens in western Ghats South-western India. In: Jain, S.K. and Rao, R.R. (eds) *An Assessment of Threatened Plants in India*. Botanical Survey of India, Howrah, India, pp. 318–322.

Purkayastha, R.P. and Chandra, A. (1976) *Indian Edible Mushrooms*. Firma K.M. Private Ltd, Calcutta, India.

Purkayastha, R.P. and Chandra, A. (1985) *Manual of Indian Edible Mushrooms*. Today and Tomorrow's Printers and Publishers, New Delhi, India.

Rattan, S.S. (1977) *The Resupinate Aphyllophorales of the North-western Himalayas*. J. Cramer, Vaduz, Liechtenstein.

Rattan, S.S. and Khurana, I.P.S. (1978) The Clavarias of the Sikkim Himalayas. *Bibliotheca Mycologica* 66. J. Cramer, Vaduz, Liechtenstein.

Rawla, G.S. and Arya, S. (1987) Some edible and poisonous agarics new to India. In: Kaul, T.N. and Kapur, B.M. (eds) *Indian Mushroom Science* II. Regional Research Laboratory, Jammu, India, pp. 375–382.

Rodgers, W.A. and Panwar, H.S. (1988) *Planning a Wild Life Protected Area Network in India*, I and II. Wild Life Institute of India, Dehra Dun, India.

Saini, S.S. and Atri, N.S. (1995) Mushroom flora of Punjab. In: Chadha, K.L. and Sharma, S.R. (eds) *Advances in Horticulture* 13, *Mushroom*. Malhotra Publishing House, New Delhi, India, pp. 374–386.

Saini, S.S., Atri, N.S. and Bhupal, M.S. (1988) North Indian Agaricales – V. *Indian Phytopathology* 41, 622–625.

Saini, S.S., Atri, N.S. and Saini, M.K. (1989) North Indian Agaricales – VI. *Journal of the Indian Botanical Society* 68, 205–208.

Sarbhoy, A.K. (1997) Biodiversity and biosystematics of Agaricales. In: Rai, R.D., Dhar, B.L. and Verma, R.N. (eds) *Advances in Mushroom Biology and Production*. Mushroom

Society of India, National Research Centre for Mushroom, Solan, Chambaghat (H.P.), India, pp. 31–37.

Sarbhoy, A.K., Aggarwal, D.K. and Varshney, J.L. (1996) *Fungi of India.* CBS Publishers and Distributors, Daryaganj, New Delhi, India.

Sathe, A.V. (1979) Agaricology in India – a review of work on Indian Agaricales. *Biovigyanam* 5, 125–130.

Sathe, A.V. (1980) *Introduction – Agaricales (Mushrooms) of South-west India.* Maharashtra Association for the Cultivation of Science, Monograph No. 1, Pune, India, pp. 1–9.

Sathe, A.V. and Deshpande, S. (1979) Agaricales of Maharashtra. In: Chattopadhyay, S.B. and Samajpati, N. (eds) *Advances in Mycology and Plant Pathology.* Oxford and IBH Publishing Co., New Delhi, India pp. 82–88.

Sathe, A.V. and Deshpande, S. (1980) Agaricales of Maharashtra State. In: *Agaricales (Mushrooms) of South-west India.* Maharashtra Association for the Cultivation of Science, Monograph No. 1, Pune, India, pp. 9–42.

Sathe, A.V. and Kulkarni, S.M. (1987) A check list of wild edible mushrooms from South-West India. In: Kaul, T.N. and Kapur, B.M. (eds) *Indian Mushroom Science* II. Regional Research Laboratory, Jammu, India, pp. 411–413.

Sathe, A.V. and Rahalkar, S.R. (1978) Agaricales from South-west India. *Biovigyanam* 3, 119–121.

Sharda, R.M. (1991) Clavarioid Homobasidiomycetes in the Himalaya: A check list. In: Khullar, S.P. and Sharma, M.P. (eds) *Himalayan Botanical Researches.* Ashish Publishing House, New Delhi, India, pp. 31–60.

Sharma, P.M. and Sidhu, D. (1991) Notes on Himalayan *Geoglossaceae.* In: Khullar, S.P. and Sharma, M.P. (eds) *Himalayan Botanical Researches.* Ashish Publishing House, New Delhi, India, pp. 13–29.

Singh, R.D. (1994) Edible mushrooms of arid zones of Rajasthan. *Abstracts of Papers Presented at the National Symposium on Mushrooms, Solan, India.* Mushroom Society of India, Solan, H.P., p. 8.

Subramanian, C.V. (1982) Tropical mycology: future needs and development. *Current Science* 51, 321–325.

Thind, K.S. (1961) *The Clavariaceae of India.* Indian Council of Agricultural Research, New Delhi, India.

Thind, K.S. (1973) The *Aphyllophorales* of India. *Indian Phytopathology* 26, 2–23.

Tivy, J. (1993) *Biogeography – a Study of Plants in the Ecosphere.* Longman Group, London.

Varma, A. and Sarbhoy, A.K. (1996) Collection and study of microbial diversity in India – a vital component in maintaining ecological balance. *Diversity* 12, 83–84.

Verma, R.N., Singh, G.B. and Mukta Singh, S. (1995) Mushroom flora of north-eastern hills. In: Chadha, K.L. and Sharma, S.R. (eds) *Advances in Horticulture* 13 *Mushroom.* Malhotra Publishing House, New Delhi, India, pp. 329–349.

Watling, R. (1978) The study of Indian Agarics. *Indian Journal of Mushrooms* 4, 30–37.

Watling, R. (1990) On the way towards a Red Data Book on British Fungi. *Transactions of the Botanical Society of Edinburgh* 45, 463–471.

Watling, R. and Abraham, S.P. (1992) Ectomycorrhizal fungi of Kashmir forests. *Mycorrhiza* 2, 81–87.

Watling, R. and Gregory, N.M. (1980) Larger fungi from Kashmir. *Nova Hedwigia* 32, 493–564.

World Wide Fund for Nature – India (1992) *WWF, India – the Conservation Plan. 1992–1993.* New Delhi, India.

Mushroom Collecting in Tanzania and Hunan (Southern China): Inherited Wisdom and Folklore of Two Different Cultures 11

M. HÄRKÖNEN

Department of Ecology and Systematics, PO Box 7, FIN-00014, University of Helsinki, Finland

Introduction

Tanzania in tropical Africa and Hunan in subtropical China both have more than half their area under forest or woodland. Both have a monsoon climate with distinct rainy seasons during which a large yield of mushrooms appears. In 1990–1999, ethnomycological field research was carried out during four visits to Tanzania (Fig. 11.1) and one to Hunan (Fig. 11.2). In both areas, outlying villages were visited, local people were accompanied to collect mushrooms, and the food prepared and eaten with them. The vernacular names and uses for the various species were also obtained. In addition, the mushroom pickers were interviewed with the help of interpreters using standard questions. When the study started very little had been written about mushroom use in Tanzania (Eichelbaum, 1906) although a large publication on the identification of East Africa agarics existed (Pegler, 1977). Currently, more and more research on the edibility of mushrooms is accumulating from neighbouring countries (e.g. Buyck, 1994; Ryvarden *et al.*, 1994) and a survey of the literature on edible and poisonous fungi of Africa south of the Sahara has now been produced (Rammeloo and Walleyn, 1993; Walleyn and Rammeloo, 1994). In China there is a long written history about eating and cultivating fungi, and their medical uses (e.g. Ying *et al.*, 1987) and a large work on the fungi of China has been published in English (Teng, 1996), and a volume on the macrofungi of Hunan in Chinese (Li *et al.*, 1993).

In neither area visited had the village people seen any mushroom identification book, so the author was able to collect local and ethnically spoken information. When the first excursion to Tanzania was planned, it was thought that

 Tropical Mycology, Vol. 1, *Macromycetes* (eds R. Watling, J.C. Frankland, A.M. Ainsworth, S. Isaac and C.H. Robinson)

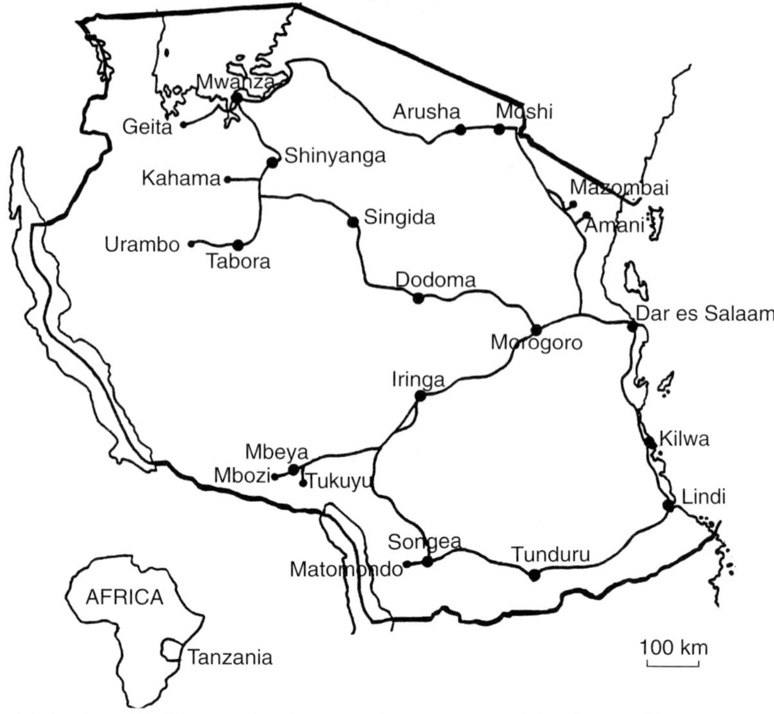

Fig. 11.1. A map of Tanzania showing the itinerary of the four collecting trips.

travelling on the frontier roads during the rainy season and communication with people of 120 vernacular languages would be the most difficult part of the survey. However, such difficulties were surmounted with the help of Leonard Mwasumbi from Dar es Salaam University. More difficult was the identification of specimens, since Tanzanians eat, and even sell in market places, mushrooms not yet scientifically described. Most of the fungal specimens collected in Tanzania currently still await identification, but the most important results concerning useful mushrooms have been published (Härkönen, 1992, 1995; Härkönen *et al.*, 1993a,b,c; 1994a, b,c; 1995; Saarimäki *et al.*, 1994; Calonge *et al.*, 1997; Karhula *et al.*, 1998). On each trip slime moulds were also collected and the number of myxomycetes reported from Tanzania now comprises 120 identified species (Härkönen and Saarimäki, 1991; 1992; Ukkola and Härkönen, 1996; Ukkola *et al.*, 1996; Ukkola, 1998a,b,c). Most of the myxomycetes collected in China have been identified, increasing the number of known species of Hunan from 14 to 77, but most of the macrofungi have been identified only to genera. For the future research plan see Härkönen (2000).

The results compiled from 100 interviews in Tanzania with people of 35 tribes, and from seven interviews with Chinese in Hunan are outlined below

Fig. 11.2. A map of Hunan showing the research areas visited in 1999 and 2000.

with the questions asked. Some small additions were made to reflect preliminary results from the second field trip to China in September 2000.

1. Do you eat mushrooms?

In Hunan everybody used mushrooms for food. In Tanzania there were large differences between tribes. The Chagga, Arusha, Meru and Maasai, for example, would not put a mushroom in their mouths, but the majority of people liked them very much. The attitude was uniform within specific tribes.

2. Who in your family collects mushrooms?

In Tanzania women and children went out mushroom hunting, but the men might bring mushrooms home if they happened to see them. In contrast, informants in Hunan said that the picking of mushrooms interested men more than women.

3. Who taught you about collecting mushrooms?

In both cultures mushroom knowledge was handed down orally from one generation to another. The most knowledgeable were elderly country people (women in Tanzania) whose families had lived in the same area for generations.

Educated people had forgotten almost everything about wild mushrooms. Vernacular mushroom lore was not taught at school and was attributed to the life of primitive food gatherers. Many urbanized Tanzanians, however, remembered with enthusiasm how as small children they used to keep an eye out for mushrooms during such activities as tending cattle, and the Chinese were ready to pay much more for wild mushrooms than for cultivated ones.

4. How many different species do you collect for food?

In Tanzania the selection of edible species was small in mountainous areas and large in miombo woodland; the highest number of species listed by one person was 28. In China only mountain areas were visited and the highest number of mushrooms counted by one man in Badakongshan was 33 species. In Zhangjiajie in Hunan only the most delicious species were collected, because it was hard work and services to the tourist industry were considered more rewarding. One method of identifying whether a *Russula* is edible is to blow on the fruit body; if the gills whistle then the fungus is considered delicious (Fig. 11.3).

5. Please list the names of mushrooms presented

The vernacular names in both areas varied greatly. The largest number, 22, was for *Termitomyces letestui* (Pat.) Heim. in Tanzania. Also, in Hunan the vernacular names for certain mushrooms were different in the three research areas visited and also differed from the names in a Chinese mushroom guide-book (Ying *et al.*, 1988; Huang 1998) with the exception of commonly cultivated mushrooms; for example, *Auricularia auricula* (Hook) Underwood and *Lentinula* (*Lentinus*) *edodes* (Berk.) Sing., whose Chinese names refer to wood's ear and fragrant mushroom respectively. The vernacular names in both cultures are illogical in the sense of scientific classification and referred to such features as colour, form, substrate, habitat or taste. For instance, all the species growing on banana litter might have a common name and, by use of compound words, species unrelated to each other might be classified in a common group. In Hunan the lichen *Umbilicaria esculenta* (Miyoshi) Minks was combined with the basidiomycete *A. auricula* through their names *Yan-er* (the ear of a rock) and *Mu-er* (the ear of wood).

6. Which species tastes best?

In this connection the informants were also asked to classify the mushrooms according to their palatability into three categories: *** delicious, ** good, * just edible. From a summary of the answers it was obvious that in Tanzania species of *Termitomyces* (Fig. 11.4) were considered the best. *T. letestui* (Pat.) Heim, *T. eurrhizus* (Berk.) Heim, *T. microcarpus* (Berk. & Br.) Heim and *T. singidensis* Saarim. & Härk. all received an average ranking of three points.

Fig. 11.3. In Hunan one way of testing whether a specimen of *Russula* is edible is to blow on its gills. This example 'whistled', so it is good for food.

There are many species of chantarelle in Tanzania, yellow, red, black and even bicoloured, and in the author's experience all were considered edible and at least two, *Cantharellus isabellinus* Heinem. and *Cantharellus platyphyllus* Heinem. as delicious.

Many species of *Amanita*, both with and without a ring, were collected for food, but only *Amanita loosii* Beeli (in our mushroom-guide, Härkönen *et al.*, 1995, with synonym *Amanita zambiana*; see Walleyn and Verbeken, 1998) received the highest praise (Fig. 11.5).

Numerous species of *Russula* and *Lactarius* occurred in the miombo woodlands and also in the mountains of Tanzania. In the mountainous area people tended to reject them, but in the miombo area some were considered edible and others non-edible. There was no apparent rule for judging edibility. They broke

Fig. 11.4. In Tanzania, specimens of *Termitomyces* are dug up with a stick to get all the pseudorrhiza. This *T. singidensis* grew in badly degraded, former woodland.

a fruit body, smelt it and had a taste and then said if it was edible or not. Among the rejected and accepted ones were both acrid and mild species. I personally tasted every specimen before preparing them for herbarium collections and, according to long experience, found that far fewer acrid species occurred in miombo woodland than in the boreal forests of northern Europe. As in my home country, Finland, the acrid species were made milder by pre-boiling them before cooking (Härkönen, 1998). Among very mild 'three-point' *Lactarius* species were: *L. cabansus* Pegler & Piearce, *L. edulis* Verbeken & Buyck, *L. xerampelinus* Karhula & Verbeken, and *L. volemoides* Karhula, which is so mild and pleasant that hunters may eat it raw in the field. Among the highest quality *Russula* species were *R. cellulata* Buyck, and *R. ciliata* Buyck. Other genera including

Fig. 11.5. Among the several edible species of *Amanita* in Tanzania, *A. loosii* is the most praised one.

species considered edible by some tribes were: *Armillaria*, *Auricularia*, *Coprinus*, *Macrolepiota*, *Pleurotus*, *Polyporus*, *Volvariella* and some ramarioid fungi. It can be generalized that people living in mountainous areas preferred wood-inhabiting fungi, and those living in miombo woodlands preferred mycorrhizal fungi, but in both areas *Termitomyces* were most valued. Most tribes considered species of *Agaricus* poisonous, although some informants knew that white people like them. With regard to boletes, Tanzanians said that even monkeys do not eat them. Although boletes are very common in miombo woodlands, we found a bolete only once, a specimen of *Strobilomyces* from the mushroom basket of a woman, and she said that it had to be boiled before cooking.

In Hunan the most appreciated 'mushroom' shown was a lichen, *Umbilicaria esculenta*, also used in Japan and Korea (Wei, 1991). It was collected from the very steep rocks in Zhangjiajie and sold in market-places at the high price of 160 yuan kg^{-1}; the other wild fresh mushrooms cost about 2–40 yuan kg^{-1}, with cultivated ones being still cheaper. In a first-class restaurant in Changsha, food made of *Ramalina* sp. and *Lobaria* sp. was also available. In the author's opinion, all three dishes were tasty.

Other species whose taste was evaluated with three 'points' by some of our Chinese informants were boletes or belonged to the genera *Pleurotus*, *Lentinus*, *Auricularia*, *Ramaria*, *Lactarius*, *Russula*, and *Amanita*. During the visit there was a good yield of the genus *Lactarius*. The species belonging to the *Lactarius deliciosus* group were most popular and sold under the common name *Tsong-jun* (meaning growing-under-pine). Mieke Verbeken (Ghent, Belgium, personal communication) has identified one of the specimens as the recently described

Lactarius subindico Verbeken & Horak (Verbeken and Horak, 2000). From preliminary identification they also included *Lactarius hatsudake* Tanaka and *L. akahatsu* Tanaka, species which are traditionally used as commercial mushrooms in Japan (Tanaka, 1890; Imazeki *et al.*, 1988). The informants preferred the bluish and greenish varieties of *Tsong-jun* (Fig. 11.6) to the orange one (*L. akahatsu*).

It was curious to a Western consumer that both Tanzanian and Chinese people eat tough, leathery species of polypores and also the rather insignificant *Schizophyllum commune* Fr. Several species of *Auricularia* have been used and also cultivated since antiquity in China and elsewhere in eastern Asia (Chandra, 1989). Most Tanzanians do not use *Auricularia* but, for instance, in the Usambara mountains they are popular and the author has recorded their

Fig. 11.6. In Hunan men like to go mushroom-hunting. Here specimens of *Lactarius* have been collected.

descriptive Swahili name *Uyoga Hindi* (= mushroom of Hindu) referring to Asia. Both *Schizophyllum* and species of *Auricularia* are considered health-promoting in China.

7. Is everyone here allowed to collect mushrooms everywhere?

In both countries there were no restrictions on collecting wild mushrooms even from other owners' plantations.

8. When and how often are the mushrooms collected?

In the Tanzanian countryside during the rainy seasons mushrooms could be collected every day. Mushroom forays were usually made early in the morning, because there was competition for the best varieties. In Hunan there were two fruiting seasons, in (March) April–June and August–September (October), but families did not go regularly to pick mushrooms because cultivated mushrooms were available throughout the year.

9. How do people prepare mushrooms for food?

In Tanzania the kitchen in the countryside was very simple and all the food was prepared on open fires between three stones. The mushrooms were washed and cut into pieces and cooked in water or oil in a kettle. Peeled onions and tomatoes were added, and sometimes also wild or cultivated vegetables but seldom with any other spices or seasonings except salt. The food was cooked for so long that the tomatoes and onions formed a thickened stew which was eaten with ugali (a type of maize-porridge), rice, cassava or cooking-bananas.

The Chinese kitchen was well equipped and had a large stove with big shallow pans. In Hunan they also began by washing the mushrooms and cutting them into pieces. Some wild species were parboiled before cooking and some less knowledgeable people pre-boiled all the wild species just for safety's sake. Usually the preparation of mushrooms began by frying some ginger and hot pepper in oil in the pan before adding the mushrooms. Then some meat such as chicken or fresh or smoked pork was added to the pan. Before taking it to the table some cooked vegetables might have been added to the serving dish. At the same meal there might be more than one dish of fungi, and the hot pot in the middle of the table often consisted of mushroom soup.

10. Who in your family prepares mushrooms for food?

That was an inappropriate question in Tanzania, because in that African society the mother always prepares the food. In Hunan men and/or women prepare food and some of the male informants said that they themselves always collected and prepared the wild mushrooms.

11. How highly do you value mushrooms as food compared with other foodstuffs such as meat, fish or vegetables?

This again was a foolish question in Hunan, because people like all those foods and usually have all of them during the same meal at least during feasting. In Tanzania most people value mushrooms very highly and, especially at the beginning of the rainy season, they are preferred to any other kind of food. The mushrooms appear just at the right time at the beginning of the rainy season when the crops of cultivated food from the previous season are almost finished and new crops have only just been planted. As food, Tanzanians consider mushrooms to be similar to meat, particularly chicken.

12. Do you preserve mushrooms?

In both areas drying was the typical method of preserving the mushrooms. In Tanzania in the miombo area there are some sunny days even during the rainy season. The mushrooms were cut into pieces and spread on a mat or an iron sheet. Some specimens of *Lactarius* were pre-boiled before drying and became very hard. The tropical sunshine is a quick dryer. However, when dried Tanzanian mushrooms were analysed in the Technical Research Centre in Finland, specimens were not considered microbiologically satisfactory for human consumption. In Hunan, dryers with sieved shelves heated from the bottom were used, and the result was excellent when estimated by eye. Wild mushrooms were mostly used only when fresh. In Daweishan where the relative humidity was above 83% the dried mushroom were also smoked after drying to prevent spoilage during storage.

13. Are mushrooms sold in the market-places in this region?

In Tanzania mushrooms are sold, fresh or dried, in market-places and on roadsides, for example, specimens of *Termitomyces*, *Cantharellus*, *Lactarius*, *Russula* and *Amanita*. Sometimes the identification appeared to be the responsibility of the buyer; we have found the poisonous *Chlorophyllum molybdites* (Meyer:Fr.) Massee among the varieties offered for sale.

In short visits to the market-places in Hunan, in addition to the cultivated *Auricularia* and *Lentinus*, only species of *Lactarius* and the lichen *U. esculenta* were seen being offered for sale. In Tanzania, mushrooms were not cultivated, but in the sisal estates visited, people came daily to collect specimens of *Coprinus* growing on composts of sisal-processing factories. In both countries the price paid for mushrooms was rather high compared, for example, with fruit or vegetables.

14. Do you use mushrooms for purposes other than for food, for instance as medicine?

In Tanzania information on medicinal use of mushrooms was scarce. In several villages mushrooms were given to mothers after childbirth to promote recovery. *T. eurrhizus* was also used mixed with some herbs as a cream for skin diseases, and *A. loosii* was considered to be a cure for stomach problems. Information on the ancient use of *T. letestui* to cure bilharzia was obtained, and a young man who collected the pseudorhizas of that species for a traditional healer was encountered, but he did not know for what purpose it was used. Luguru people used *Auricularia* species for removing ear wax. The fruit body was warmed by a fire and the jelly-like content was pressed into the ear. The wax could then be removed from the ear. Species of hard polypores, including *Ganoderma* were used widely for such problems as toothache, coughs and animal diseases. In several places puffballs were used to sedate bees. A puffball was attached to a stick and set on fire. It produced a lot of smoke and could be taken on the stick to sedate a hive of bees.

In China the use of fungi for medicine has a long history. The earliest book on medical substances in China from the 1st century BC included fungi, and a modern book on medicinal fungi included descriptions of as many as 272 fungal species (Ying *et al.*, 1987). Mushrooms are used not only to cure diseases, but their regular use as food is also considered to provide a healthy and balanced diet. So the functional foods which have become so fashionable in Western countries have been used since antiquity in China. Modern Chinese pharmacological research has discovered more and more substances in fungi which promote health and cure diseases such as sarcoma, carcinoma and hepatitis. Most informants, however, used mushrooms only for food. They considered mushrooms to be health-promoting, and it was customary to bring mushrooms as a present when visiting friends.

15. Is there a traditional healer in your village who uses mushrooms? How are they used?

Some traditional healers met in Tanzania were friendly but not willing to describe their medicines. Such matters were secret, since the medicines lose their power if their composition is known to outsiders.

In China the opposite situation occurred. In Badagongshan the local herbalist, Tao Jie-wen, presented several specimens and proudly said how and for what purpose these medicinal fungi were used. Several species of polypores were used; for instance *Polyporus tubaeformis* (P. Karst.) Ryvarden & Gilb. could be used against headaches. It was cleaned, dried and powdered, and the powder was mixed with water and drunk. It is also good for digestion, when washed, dried, powdered and diluted with alcohol or warm water. Some species of *Phellinus*, e.g. *Phellinus baumii* Pil., were powdered and mixed with vegetable oil to use against skin diseases such as scabs and small wounds, or the powder was placed directly on to the skin. *Fomes fomentarius* (L.: Fr.) Fr. was used against kidney disease. It was soaked in alcohol or used with other medicines. Food prepared with *Auricularia* is good for the digestion. He also brought out long, white, string-like mycelial cords, collected from vegetation in the forest and used against hepatic cirrhosis. It was chewed as it was or used with other medicaments. Some white radial mycelial fans on wood were used against neurasthenia. The mycelium was eaten raw and alcohol drunk with it. The medicine-man also presented several specimens of lichens used for medical purposes such as healing of wounds or burned skin.

Considered best of all medical fungi in China is *Ganoderma lucidum* (M.A. Curtis: Fr.) P. Karsten. It has been said to cure almost every ailment; for example headache, sleeplessness, poor appetite, 'womens' troubles' and depression. It is used diluted in water like tea, or with alcohol. In addition, it brings good luck. The Chinese have learned how to cultivate *Ganoderma* and it was offered for sale in shops and markets, for instance in the Zhanjiajie Nature Reserve. Needless to say the wild specimens were considered much more potent and were more expensive than the cultivated. The bigger and more branched the fruit body the more desired it was. In Zangjiajie, a man who had found an enormous fruit body of peculiar shape earned his living by showing it to tourists.

In Tanzania some peculiar fungi were used as amulets. For instance, near Mbeya a young man was seen carrying a hard fruit body of a tall *Xylaria*. When he met his enemy he just had to take it from his pocket and scratch his head with it, and all antagonism was forgotten!

16. Are there poisonous mushrooms in this region?

In both countries people were very well aware of the existence of poisonous mushrooms.

17. Have misidentifications occurred, leading to mushroom poisonings?

Mushroom poisonings, even fatal ones, were reported in both countries. In Tanzania mushroom poisonings for the purpose of killing, for instance an undesirable daughter-in-law, were recalled.

18. How do you recognize a poisonous mushroom?

In general, in both countries informants seemed to be uninterested in knowing or naming the poisonous species. A collective name was used for them. When asked, some tried to describe some general characters of poisonous species. In Tanzania, there was a common belief that if the gills of the fungus turn dark, as in the genus *Agaricus*, the species is very poisonous. In China it was believed that all the species with bright colours (e.g. *Aseroë rubra* Labill., *Calostoma cinnabarina* Desv. and *Dictyophora multicolor* Berk & Br.) are poisonous. In Daweishan the following dangerous rule was used for unknown pleasant-looking mushrooms: collect some sprouts of *Juncus decipiens* and peel their base. Put the peeled bases in boiling water with mushrooms. If they change colour, the mushrooms are poisonous. Most people in both countries realized that poisonous mushrooms have no common character. Mushroom poisoning was avoided by carefully identifying the edible ones as taught by their parents. That was commendable, but a double check would add to safety and people should be taught what the most dangerous mushrooms look like; indeed there is a popularized guide on poisonous mushrooms available in China (Anonymous, 1988). Both in China and in Tanzania plantations of exotic trees are common. Mycorrhizal mushroom species unknown to the inhabitants might be brought into the area with the tree seedlings. This must have happened in the monotonous plantations of *Pinus patula* in Mufindi and Mafinga in 1994, where the needle layers of the plantations were full of red fly agaric, *Amanita muscaria* (L.: Fr.) Hooker, common in Europe and Asia but not indigenous to tropical Africa. In a local hospital some victims of mushroom poisoning by that species verified that they had thought they had consumed a tasty edible red *Amanita, A. tanzanica* Häk. & Saarim., *Wigwingwi* in Hehe language (Fig. 11.7). The symptoms of strong hallucinations were typical of *A. muscaria* (Fig. 11.8), leading to the unexpected death of one of the victims.

19. Do you know any beliefs, stories or fairytales about mushrooms?

In Tanzania several stories and even songs about mushrooms have been collected (see Härkönen *et al.*, 1995). Some are nice fables, some are mythical ideas about the powers of some mushroom species and some have an ecological moral. In Hunan only some were heard; one is as follows. In the autumn a wounded crane flew towards the south and its blood was dropping into a pine forest. From this blood *Tsong-jun* (*L. hatsudake*) with reddish latex milk grew. The other story was connected to a vary rare mushroom *Jiangnü-Jun* which grew from the tears of a beautiful woman, Jiangnü, who wept desperately because her husband was taken for years to build the Great Wall of China.

Fig. 11.7. In Tanzania the indigenous *Amanita tanzanica* is a good edible mushroom.

Discussion

When collecting vernacular information on the use of mushrooms in the frontier areas of Tanzania and Hunan, the richness of orally inherited knowledge was seen, and the surface of it has only been scratched. Let us hope that more people, and especially indigenous mycologists, will direct their enthusiasm to the field of ethnomycology. It would be valuable ecologically, economically and culturally. But there needs to be haste, because urbanization and the modern way of living results in societies forgetting their long oral traditions.

Fig. 11.8. The introduced *Amanita muscaria* resembles *Amanita tanzanica,* especially when it loses the white spots of its cap as the fruit-bodies push through the needle-cover under exotic pines, so causing misidentification and poisoning.

References

Anonymous (1988) *Poisonous Mushrooms* (in Chinese). Fungal section of the Institute of Microbiology of the Chinese Academy of Sciences, Science Press, Beijing, China.

Buyck, B. (1994) Ubwoba: Les champignons comestibles de l'Ouest Burundi. *Administration Générale de la Cooperation au Développement, Publication Agricole* 34, 1–123.

Calonge, F., Härkönen, M., Saarimäki, T. and Mwasumbi, L. (1997) Tanzanian mushrooms and their uses 5. Some notes on Gasteromycetes. *Karstenia* 37, 3–10.

Chandra, A. (1989) *Elsevier's Dictionary of Edible Mushrooms.* Elsevier Science, The Netherlands.

Eichelbaum, F. (1906) Beiträge zur Kenntnis der Pilzflora des Ostusambaragebirges. *Verhandlungen des Naturwissenschaftlichen Vereins in Hamburg* 14, 1–92.

Härkönen, M. (1992) Wild mushrooms, a delicacy in Tanzania. *Universitas Helsingiensis* 11, 29–31.

Härkönen, M. (1995) An ethnomycological approach to Tanzanian species of *Amanita. Symbolae Botanicae Upsalienses* 30, 145–151.

Härkönen, M. (1998) Uses of mushrooms by Finns and Karelians. *International Journal of Circumpolar Health* 1/1998, 40–55.

Härkönen, M. (2000) The fabulous forests of southern China as a cooperative field of exploration. *Universitas Helsingiensis* 19, 20–22.

Härkönen, M. and Saarimäki, T. (1991) Tanzanian myxomycetes. First survey. *Karstenia* 31, 31–54.

Härkönen, M. and Saarimäki, T. (1992) Tanzanian myxomycetes. First survey (addition). *Karstenia* 32, 6.

Härkönen, M., Buyck, B., Saarimäki, T. and Mwasumbi, L. (1993a) Tanzanian mushrooms and their uses 1. *Russula. Karstenia* 33, 11–50.

Härkönen, M., Saarimäki, T. and Mwasumbi, L. (1993b) Tanzanian mushrooms and their uses 2. An edible species of *Coprinus* section *Lanatuli. Karstenia* 33, 51–59.

Härkönen, M., Saarimäki, T., Mwasumbi, L. and Niemelä, T. (1993c) Collection of the Tanzanian mushroom heritage as a form of developmental cooperation between the universities of Helsinki and Dar es Salaam. *Aquilo Series of Botany* 31, 99–105.

Härkönen, M., Saarimäki, T. and Mwasumbi, L. (1994a). Setting up a research project on Tanzanian mushrooms and their uses. *Proceedings of the 13th Plenary Meeting of AETFAT, Zomba, Malawi, 2–11 April 1991* 1, 735–748.

Härkönen, M., Saarimäki, T. and Mwasumbi, L. (1994b) Tanzanian mushrooms and their uses 4. Some reddish edible and poisonous *Amanita* species. *Karstenia* 34, 47–60.

Härkönen, M., Saarimäki, T. and Mwasumbi, L. (1994c) Edible and poisonous mushrooms of Tanzania. *African Journal of Mycology and Biotechnology* 2, 99–123.

Härkönen, M., Saarimäki, T. and Mwasumbi, L. (1995) Edible mushrooms of Tanzania. *Karstenia* 35, Suppl., 1–92.

Huang, N. (1998) *Coloured Illustrations of Macrofungi (Mushrooms) of China* (in Chinese). China Agricultural Press, Beijing, China.

Imazeki, R., Otani, Y. and Hongo, T. (1988) *Fungi of Japan* (in Japanese). Yama-Kei Publishers, Tokyo.

Karhula, P., Härkönen, M., Saarimäki, T., Verbeken, A. and Mwasumbi, L. (1998) Tanzanian mushrooms and their uses 6. *Lactarius. Karstenia* 38, 49–69.

Li, J., Hu, X. and Peng, Y. (1993) *Macrofungus flora of Hunan* (in Chinese). Hunan Normal University Press, Changsha.

Pegler, D.N. (1977) A preliminary agaric flora of East Africa. *Kew Bulletin Additional Series* 6, 1–615.

Rammeloo, J. and Walleyn, R. (1993) The edible fungi of Africa south of the Sahara: a literature survey. *Scripta Botanica Belgica* 5, 1–62.

Ryvarden, L., Piearce, G.D. and Masuka, A.J. (1994) *An Introduction to the Larger Fungi of South Central Africa*. Baobab Books, Harare.

Saarimäki, T., Härkönen, M. and Mwasumbi, L. (1994) Tanzanian mushrooms and their uses 3. *Termitomyces singidensis*, sp. nov. *Karstenia* 34, 13–20.

Tanaka, N. (1890) On Hatsudake and Akahatsu, two species of Japanese edible fungi. *Botanical Magazine Tokyo* 4, 2–7.

Teng, S.C. (1996) *Fungi of China*. Mycotaxon, Ltd, Ithaca, New York.

Ukkola, T. (1998a) Tanzanian myxomycetes to the end of 1995. *Publications in Botany from the University of Helsinki* 27, 1–45.

Ukkola, T. (1998b) Some myxomycetes from Dar es Salaam (Tanzania) developed in moist chamber cultures. *Karstenia* 38, 27–36.

Ukkola, T. (1998c) Myxomycetes of the Usambara Mountains, NE Tanzania. *Acta Botanica Fennica* 160, 1–37.

Ukkola, T. and Härkönen, M. (1996) Revision of *Physarum pezizoideum* var. *pezizoideum* and var. *microsporum* (Myxomycetes). *Karstenia* 36, 41–46.

Ukkola, T., Härkönen, M. and Saarimäki, T. (1996) Tanzanian myxomycetes. Second survey. *Karstenia* 36, 51–77.

Verbeken, M. and Horak, E. (2000) *Lactarius* (Basidiomycota) in Papua New Guinea 2. Species in tropical-montane rainforests. *Australian Systematic Botany* 13(5), 649–707.

Walleyn, R. and Rammeloo, J. (1994) The poisonous and useful fungi of Africa south of the Sahara: a literature survey. *Scripta Botanica Belgica* 10, 1–56.

Walleyn, R. and Verbeken, A. (1998) Notes on genus *Amanita* in Sub-Saharan Africa. *Belgian Journal of Botany* 131, 156–161.

Wei, J.C. (1991) *An Enumeration of Lichens in China.* International Academic Publishers, Beijing.

Ying, J., Mao, X., Ma, Q., Zong, Y. and Wen, H. (1987) *Icones of Medical Fungi From China.* Science Press, Beijing.

Ying, J., Zhao, J., Ma, Q., Xu, L. and Zong, Y. (1988) *Edible Mushrooms* (in Chinese). Science Press, Beijing.

Impact of Developmental, Physiological and Environmental Studies on the Commercial Cultivation of Mushrooms

12

D. Moore[1] and S.W. Chiu[2]

[1]*School of Biological Sciences, University of Manchester, Oxford Road, Manchester M13 9PT, UK;* [2]*Department of Biology, The Chinese University of Hong Kong, Shatin, Hong Kong SAR, China*

Introduction

Fungi are an ideal food because crude protein typically represents 20–30% of fungal dry matter and contains all of the amino acids which are essential to human and animal nutrition. Fungal biomass is digested to leave their chitinous walls as a source of dietary fibre. Fungi contain B vitamins and are characteristically low in fat. An extremely important attribute of all fungal food is that it is virtually free of cholesterol. Fungal protein foods can compete successfully with animal protein foods like meat on health grounds. Since fungal foods can be produced readily, in principle, using waste products as substrates, fungal foods should also be able to compete successfully on grounds of primary cost.

Agaricus bisporus (J.E. Lange) Imbach is by far the most commonly cultivated mushroom, accounting for over 70% of total global mushroom production in the mid-1970s (Moore and Chiu, 2001). Currently, it accounts for something closer to 30% even though production tonnage has more than doubled in the intervening years. A major change during the last quarter of the 20th century was the increasing interest shown in so-called 'exotic' mushrooms. Fresh *Lentinula* (*shiang-gu* in Chinese or *shiitake* in Japanese) and *Pleurotus* (oyster mushrooms) are routinely found alongside *Agaricus* in supermarkets around the world.

One reason for the remarkable increases seen in production of certain mushrooms has been the use of industrial waste products as substrates. Oyster mushrooms are easily grown on a variety of agro-industrial wastes, including cotton wastes. Similarly, while the paddy straw mushroom (*Volvariella volvacea*

(Bull.: Fr.) Singer) is traditionally grown in South-east Asia on rice straw, it too can be grown on cotton waste. Indeed, cotton waste (generated by the textile and garment industries) gives higher yields and is also more widely available than is rice straw, so it is far cheaper (rice straw becomes expensive because of the cost of transport). Typically 80–90% of the total biomass of agricultural production is discarded as waste. This is an unacceptable loss of primary production when many mushrooms could be grown on agricultural residues, converting wastes into food, animal feed, pharmaceuticals and other products (Chang, 1998).

This chapter briefly illustrates the wide variety of recent research on the physiology, genetics and morphogenesis of fungi relevant to the commercial cultivation of mushrooms. Some of this research remains to be exploited.

Mushroom Cultivation can be a Profitable and Environmentally Friendly Process

Cultivation of an *Agaricus* crop depends on composted plant litter. Similar approaches have been developed for oyster and paddy straw mushrooms in the Far East, although in the Chinese tradition the most important mushroom crop (*Lentinula*) is cultivated on wood logs. Indeed, this traditional technique is still the most frequently used method in China over a growing region which covers a territory about equal to the entire land area of the European Union. Continuation of the traditional use of locally cut logs is likely to devastate the hill forests. For this and other reasons more industrial approaches are being applied to shiitake growing. Hardwood chips and sawdust packed into polythene bags as 'artificial logs' provide a highly productive alternative to natural logs, and the cultivation can be done in houses (which may only be plastic-covered enclosures) in which climate control allows year-round production.

Improved use of substrates depends on a better understanding of the growth physiology of the fungus. A particularly interesting series of experiments on protein utilization by basidiomycete fungi was carried out by Kalisz *et al.* (1986). Protein is probably the most abundant nitrogen source available to organisms in compost in the form of lignoprotein, microbial protein and plant protein. *A. bisporus* has been shown to be able to degrade dead bacteria and to utilize them as the sole source of carbon, nitrogen and phosphorus (Fermor and Wood, 1981), and all three species studied by Kalisz *et al.* (1986), *A. bisporus*, *Coprinus cinereus* (Schaeff.: Fr.) S.F. Gray and *V. volvacea*, were able to use protein as the sole source of carbon, nitrogen and sulphur. Furthermore, protein was utilized as efficiently as was glucose when provided as a sole source of carbon. Supplied together as carbon sources, both protein and glucose were utilized more rapidly, and growth was greater, than when either protein or glucose was supplied separately as a sole source of carbon. Extracellular proteinase of *Agaricus*, *Coprinus* and *Volvariella* was not subject to catabolite repression by glucose or ammonium regulation as described for the ascomycetes *Aspergillus* and *Neurospora*. Rather, basidiomycete proteinase activity is regulated primarily by induction, although it is subject to sulphur-, carbon- and nitrogen-catabolite derepression to different extents in the different organisms (Kalisz *et al.*, 1989). Another important point is that, *in vitro*, loss of substrate protein from the medium was detectable before extracellular proteinase activity was evident in the medium. Thus, proteinase activity is initially localized to the hyphal wall in basidiomycetes; the enzymes are released only when proteinaceous substrates approach exhaustion. These 'basidiomycete types' of regulation and localization are presumably adaptations to the plant litter habitat. Understanding them should help in the design of artificial composts and in the choice of compost supplements.

Despite the fact that the natural habitat of *Lentinula edodes* (Berk.) Pegler is the timber of felled trees, it can be grown effectively, both as mycelium and fruit bodies, in liquid media *in vitro*. Tan and Moore (1992) found that several commercial strains formed good mycelial growth in defined liquid medium in both stationary and shake flask cultures. Supplementation of flask cultures with vermiculite promoted growth and fruiting and the best fruiting *in vitro* was obtained by inoculation on to solid supports. This research demonstrated that liquid inoculum can be produced in quantity by homogenization with no adverse effect on inoculum potential, and it identified convenient media and growth conditions which permitted good yields of mycelial biomass over both long and short incubation periods.

Many agro-industrial wastes are of no apparent use, and are frequently difficult to handle and even more difficult to dispose of or treat in an environmentally acceptable way. They may even be harmful to the environment and potentially hazardous because some of them, such as olive oil waste waters and some cotton residues, are toxic to plants and animals. Disposal of such wastes in soil, water or into urban sewers is illegal in most countries and they can be major pollutants. Many commonly cultivated mushroom fungi, especially *Pleurotus* and *Lentinula* species, are white rot fungi showing high efficiency in lignocellulolytic degradation of a wide range of plant litter (Kerem *et al.*, 1992) which they selectively delignify (Moyson and Verachtert, 1991; Ortega *et al.*, 1992; Hadar *et al.*, 1993). Moreover, they grow well when conventional substrates are supplemented with solutions containing high amounts of phenols and/or tannins (Tomati *et al.*, 1991; Upadhyay and Hofrichter, 1993) and even oil in oil refinery waste waters (S.W. Chiu, Chinese University of Hong Kong, unpublished). Such fungi are able to bioconvert useless agricultural waste into the following:

1. Edible fruit bodies of high organoleptic properties and nutritive value (Chorvathova *et al.*, 1993; Zhang *et al.*, 1994; Bobek *et al.*, 1994). In terms of mushroom cultivation, *Pleurotus* species run second in annual world commercial production, representing the greatest increase (approximately fivefold) over recent years to meet market demands (Chang and Miles, 1991). The black oak mushroom, *L. edodes*, was first cultivated in China and is the most popular dried mushroom in Hong Kong, China and Japan. It holds the third place in the world production table (Chang, 1993). In addition to its unique flavour, a commercially available polysaccharide product extracted from this mushroom named lentinan is a host defence potentiator, improving immunological defence mechanisms against cancer and other diseases (Chihara, 1993). Thus, both mushrooms are important and established commercial products.

2. Good quality fodder. Enriching the waste with fungal protein through the growth of the mycelium improves digestibility by preferentially removing lignin and hemicellulose, leaving cellulose mostly intact as an energy source for ruminants (Martinez *et al.*, 1991; Tripathi and Yadar, 1992).

3. Soil conditioners/fertilizers that benefit overall soil fertility and stability and improve vegetable yield (Saiz-Jimenez and Gomez-Alarcon, 1986; Balis *et al.*, 1991; Flouri *et al.*, 1995).

These advantages have made disposal of an abundant waste coupled with production of a mushroom cash crop a popular model in recent years. *Pleurotus* spp. in particular grow readily on so many lignocellulose agricultural wastes that it becomes an attractive notion to use the fungus to digest the waste while producing crops of mushrooms. *Pleurotus* cultivation may even aid removal of pollutants from landfill or contaminated waste sites because the spent compost contains populations of microorganisms which, since they can digest the natural phenolic components of lignin can also break down chemicals such as polychlorinated phenols. Chiu *et al.* (1998a) demonstrated that the mushroom substrate left after harvesting oyster mushrooms is able to remove the biocide pentachlorophenol (PCP) more effectively than any fungal mycelium. PCP has been the most heavily used pesticide throughout the world, but it is very persistent in the natural environment and a common cause of soil contamination. Bacteria and fungi rapidly, and completely, digested PCP over a wide range of initial concentrations, whereas fungal mycelial incubations left a variety of breakdown products, some of which were also toxic. In many countries spent mushroom substrates are often discarded as wastes. Using them in landfill or contaminated sites would combine soil conditioning with degradation of organopollutants as a prospective bioremediation strategy.

Care must be exercised, however, because *Pleurotus*, like other mushrooms, accumulates metal ions in the basidiome. Wastes gathered from industrial sources for use in mushroom compost may be contaminated by heavy metals to an extent sufficient to render the crop unsuitable for consumption. For example, Chiu *et al.* (1998b) showed that cadmium could be accumulated in *Pleurotus* fruit bodies to such high levels that a single modest serving of mushrooms could cause the consumer to exceed the tolerable food limit recommended for a full week of intake of this metal.

Metabolism and Morphogenesis

In vitro studies of glucose catabolic pathways in *L. edodes* have used radiorespirometry and enzyme analyses (Tan and Moore, 1995). The ^{14}C-1/^{14}C-6 ratios in CO_2 respired from specifically labelled glucose fed to *L. edodes* tissues grown on a chemically defined medium ranged from 2.5 (vegetative mycelium) to 14.9 (young lamellae). This mirrored the relative activity of the pentose phosphate pathway (PPP), very high in the basidiome but low in the mycelium. The highest ratio was recorded in tissues which are biosynthetically most active (young lamellae), which require the reducing power of NADPH generated through the PPP. Extensive conversion of ^{14}C-3,4-labelled glucose to $^{14}CO_2$ in the mycelium underlined the important role of the Embden–Meyerhof–Parnas pathway (EMP) in that tissue. A ^{14}C-1/^{14}C-6 ratio of 14.8 for young lamellae grown on woodchips indicates that the growth medium did not influence the pathway used in the basidiome and showed that the *in vitro* data were truly representative. Ratios for the young pileipellis and stipe were 8.0 and 10.4 respectively. Tan and Moore (1995) used *Coprinus cinereus* as a comparison and found a ^{14}C-1/^{14}C-6 ratio of 3.6 in *C. cinereus* gills, confirming the comparatively lesser importance of the PPP in this organism as suggested by much earlier enzymological analyses (Moore and Ewaze, 1976). Activity of PPP enzymes (glucose 6-phosphate dehydrogenase (G6PDH) and 6-phosphogluconate dehydrogenase) was three times as high in the basidiome compared with the mycelium, highest activity being in the young pileus (Tan and Moore, 1995). EMP enzymes (fructose 1,6-bisphosphate aldolase (ALD) and glucose 6-phosphate isomerase (PHI)) were more active in mycelium than in the basidiome, and also more active than PPP enzymes within the mycelium itself. Within the basidiome, specific activities of EMP and PPP enzymes were about the same. For comparison, enzymic activity in the mature pileus of *C. cinereus* resembled the pattern in the mycelium rather than that of the basidiome of *L. edodes*. EMP enzymes were very much more active than PPP in *C. cinereus*. *Pleurotus pulmonarius* (Fr.) Kummer (often misnamed *Pleurotus sajor-caju*, Chiu *et al.*, 1998b) also had the PPP as the predominant pathway in hexose metabolism of basidiome tissues (Chiu and Moore, 1999a). Specific activity of the EMP enzyme PHI was twofold higher than that of the PPP enzyme G6PDH in vegetative mycelium, but in basidiomes, specific activity of G6PDH was at least 12-fold greater than that of PHI (Chiu and Moore, 1999a).

A positive correlation between high PPP activity and accumulation of large amounts of mannitol in the basidiome is notable. Quantitative determination of mannitol, using gas–liquid chromatography showed that mycelia of *L. edodes* had a low level of mannitol (about 1% on a dry weight basis) compared with the basidiome stipe and pileus (20–30%). The highest level of mannitol was observed in the pileus of *A. bisporus* (close to 50%), while no mannitol was detectable in the pileus of *C. cinereus* (Tan and Moore, 1994). The contrasts between *C. cinereus* on the one hand and *L. edodes* and *A. bisporus* on the other, serve to emphasize the correlation between mannitol accumulation and high activity of the PPP. Mannitol possibly acts as a store of reducing power and osmoregulatory agent, controlling the influx of water necessary for pileus expansion, although it must be emphasized that basidiome growth in *L. edodes* and *Pleurotus pulmonarius* occurs by hyphal multiplication rather than by hyphal inflation evident in *Agaricus* (see discussion in Tan and Moore, 1994). If the PPP does affect regulation of water influx (through its influence on mannitol content) there could be commercial implications in researching ways to manipulate the PPP in the basidiome (and, consequently, its water content) since 90% of the fresh mushroom consists of water. For commercial production, a mushroom crop is harvested at a particular stage of development to suit the size and shape requirements of the market. It may be possible to manipulate water content of the crop which may shorten the time taken to reach the appropriate morphological stage.

A completely different strategy for linking metabolism to morphogenesis, using urea rather than mannitol as an osmotic metabolite, was evident from earlier work on the enzymology of basidiome development in *C. cinereus*. This emerged from the discovery that the NADP linked glutamate dehydrogenase (NADP-GDH) occurs at high activity in the basidiome pileus, but is absent from the stipe and parental mycelium (Stewart and Moore, 1974). Further work showed that there are several enzymes (glutamine synthetase, ornithine acetyl transferase and ornithine carbamyl transferase) that similarly increase in activity in the developing basidiome pileus. Interestingly, urease activity is virtually absent from the pileus (though constitutive in stipe and mycelia). The whole family of enzymes shows coordinated regulation (Ewaze *et al.*, 1978) and forms a pathway which leads to formation and accumulation of urea. Regulation *in vivo* depends on accumulation of acetyl CoA and absence of ammonia (Moore, 1981). NADP-linked GDH activity is high whether conditions demand amination or deamination, and kinetic analyses also supported the view that the NAD-GDH is equally able to aminate or deaminate (Al-Gharawi and Moore, 1977). Consequently NADP-GDH function is unrelated to general metabolism, but is specifically concerned with cell differentiation. Accumulation of urea has been demonstrated using isotopically labelled precursors (Ewaze *et al.*, 1978): the quantity of urea contained in the pileus increases more than four times (on a dry weight basis) as primordia mature to the stage of spore release. On the other hand, the concentration of urea (on a fresh weight basis) remained essentially unchanged during basidiome development; the specifically amplified enzyme activities drive the urea cycle to form metabolites (mainly urea) which promote the osmotic uptake of water which is needed for the cell expansion involved in basidiome maturation. Interestingly, in *P. pulmonarius* both the PPP and the urea cycle may generate osmoregulators. The PPP predominated in basidiome tissues and urea content remained high and unchanged as cap tissues developed, although it decreased to zero at maturity in the stipe (Chiu and Moore, 1999a).

Assessing the Predictability of Mushroom Morphogenesis

The mushroom basidiome (or its equivalent where the cultivated fungus is not a basidiomycete) *is* the crop. Most interest in morphogenesis has concentrated on *Coprinus* (Moore, 1998), although Umar and van Griensven (1995, 1997a, b, c, 1998, 1999) have taken up the challenge of *Agaricus*, and colleagues in Hong Kong and Mexico are now studying *Lentinula*, *Ganoderma* and *Pleurotus*. Unfortunately we are still woefully ignorant of the most basic structural and developmental details of basidiomycete fruit bodies although these are the highest expressions of cellular and tissue organization in the fungal kingdom. One of the most fundamental biological problems which remains unresolved is the way in which cell differentiation is coordinated in the formation of organized tissues. Burnett (1968, p. 144) identified hyphal growth, branching and aggregation as the three fundamental factors involved in fungal differentiation. Thirty years later we are still hypothesizing about hyphal growth and we know virtually nothing about branching and aggregation. Models have been proposed which account for the development of complex morphologies in animals and plants. Growth factors acting as activators and inhibitors are assumed to diffuse through the tissues regulating cell differentiation. Distribution of morphogens depends on adequate communications within the tissue; and this must extend over many cell diameters. Higher plants and animals are well provided with avenues for such communication, via cell processes, desmosomes, gap junctions, plasmodesmata and the like, but electron micrographs show considerable space between fungal cells of young mushroom tissues. Indeed, at about the time that the morphogenetic prepatterning must be taking place, hyphal cells are quite distant and there is no evidence for any lateral cytoplasmic contact between neighbours. However, there is evidence of lateral communication of signals, at least during hymenium differentiation in *Coprinus*. As noted, NADP-GDH occurs at high activity in the basidiome pileus, but is absent from the stipe and parental mycelium. This enzyme can be specifically detected in histological preparations of living tissues using a tetrazolium staining procedure (Elhiti *et al.*, 1979). Application of this technique has shown that an overall increase in enzyme activity in the pileus does not occur through a uniform increase in each constituent cell. Rather, at early stages a scattering of cells show high activity, and it is the proportion of cells showing the enzyme activity which increases as the tissues mature. These scattered cells first appear in narrow stripes across the gill and as the tissues mature the stripes become wider until eventually all the hymenial cells show activity of this enzyme (Elhiti *et al.*, 1979). Evidently, there is a channel for lateral communication, but the relationships between adjacent cells are known in only the vaguest way and evidence for the existence of chemical growth factors or hormones is confused and inconclusive (Novak Frazer, 1996).

Most mycologists will be aware of the description of tissue construction called hyphal analysis introduced by Corner (1932a, b, 1966; Redhead, 1987).

Hyphal analysis has been almost entirely descriptive, and its taxonomic importance is immense (Pegler, 1996). The only *quantitative* study has been done by Hammad *et al.* (1993a, b), who showed that enumerating cell types at different stages of development in the fruit bodies of *C. cinereus* was a powerful way of revealing how basidiome structure emerges during morphogenesis as a result of changes in hyphal type and distribution. Counting and measuring hyphal compartments in different regions of fruit bodies at different stages of development reveal the mechanical generation of the final form of the basidiome. The patterns revealed must be organized by signalling molecules, so these studies raise crucial questions about the nature of such signalling molecule(s), their transduction pathways and the responses they elicit. Attempts are under way to exploit further the numerical approach, coupling it to computer-aided image analysis with the aim of establishing mathematical models describing mushroom morphogenesis. The 'virtual mushroom' may sound rather fanciful, but it is likely to be of value to industry. Relatively simple measurements of how dimensions change during morphogenesis have been used to define the 'normal' mushroom for the *A. bisporus* crop (Flegg, 1996), and image analysis of shape, form and colour of *A. bisporus* can be related to crop development (van Loon *et al.*, 1995) in ways that contribute to defining control programmes for automated harvesting.

The Question of Biodiversity

Biodiversity is the natural resource which can be exploited for breeding new cultivars to satisfy the expanding and diversifying demands of consumers. *L. edodes* has been cultivated in China for over 800 years, but the cultivars used for commercial cultivation throughout China have been found to be genetically homogeneous (Chiu *et al.*, 1996). In addition, traditional production methods pose a number of environmental problems. The traditional log-pile cultivation method is still the one most frequently used. For this, locally felled logs of oak, chestnut, hornbeam, maple and other trees over 10 cm diameter (probably *c.* 20–30 years old) and 1.5–2 m long are normally cut in spring or autumn of each year. Up to 100,000 trees must be felled every year just to maintain current production levels. Traditional usage of natural wood logs has been pursued to the extent that as availability of mature trees has declined, attention has turned to younger trees and other tree species. This, combined with other demands for timber, has contributed to a loss of 87% of the native forests in China (Anonymous, 1997). China now faces the problem that the rate of deforestation is much greater than the rate of reforestation in the remaining 13% forest cover (Mackinnon *et al.*, 1996; Loh *et al.*, 1999). There are planting regulations and prohibitions on felling of young trees, but these are difficult to monitor and 'conservation awareness' is especially low among those living in remote mountainous regions.

Recently the authors have investigated the population biology of *L. edodes* by examining natural populations in several provinces of China (Chiu *et al.*,

1998c, 1999a,b). The research covers a geographical area which is *c.* 1700 km north to south and 700 km west to east, but includes detailed surveys down to individual logs, includes phenotypes varying from morphology and palatability to DNA sequences. If mushroom farming is to become a sustainable industry, causing minimal disturbance to the natural habitats, alternative resources to logs collected from hillside forests must be promoted. Also, since a reproductive cycle for *L. edodes* takes over 6 months using artificial logs, fast fruiters are desirable, and pathogen resistance is an advantage in mass production. To focus our analysis on a sustainable mushroom cultivation industry, isolates from the field were tested for tolerance to a common bacterial pathogen, and for fruiting ability in indoor cultivation on artificial logs made of sawdust wastes from furniture manufacture.

Although a very limited gene pool is exploited in the cultivated strains in China, enormous biodiversity occurs in the wild. Analysis of local populations reveals that *L. edodes* strains show strong somatic incompatibility reactions and individual territories can be small (a few hundred mm^3). The widespread nature of the species and absence of other means of dispersal indicated that basidiospores are the major, even only, method of natural distribution. This is why harvesting *after* initiation of basidiospore release places the natural gene pool under threat of contamination between cultivated and natural populations of *L. edodes*. Protection of the natural environment is still the best strategy for conserving the biodiversity of this important commercial resource, but collection of wild strains for preservation in a culture collection would directly conserve the wild germplasm. Making a gene bank readily accessible to the public and industry would also generate a commercial resource for exploitation in both cultivation and breeding programmes. It might also reduce the pressure caused by non-professional collecting in the wild. The authors' studies suggested that a move to indoor cultivation, less dependence on multispore spawns and exploitation of a wider range of natural genotypes would better safeguard both cultivated and natural populations of the fungus and avoid denuding hillsides of mature trees.

The concept of sustainable management is too novel for the mushroom farmers of a developing country. However, villages in China are accustomed to working cooperatively to establish a shared facility, and the advantages of indoor cultivation to the farmer (consistency of yield, much shorter production cycle, use of solid industrial/agricultural wastes in the substratum) can be readily appreciated. An essential step is to educate the public and introduce these ideas. Morphological and genetic plasticity is another aspect of biodiversity which is especially evident in *Volvariella* spp. Studies of *Volvariella bombycina* (Schaeff.: Fr.) Singer have revealed enormous phenotypic plasticity (Chiu *et al.*, 1989). A wide range of spontaneous basidiome peculiarities, which are not disease symptoms, are within the normal range of development. These can be amplified and encouraged by environmental stress, such as desiccation. In the more widely cultivated *V. volvacea* most commercial spawns are heterokaryotic although the organism is homothallic. Thus, a mycelium grown from a single (haploid) basidiospore is able to produce fruit bodies and this may be seen as a good means of selecting promising commercial genotypes. However, selfed haploids of *V. volvacea* continue to segregate genetically diverse progeny through several generations of selfing (Chiu and Moore, 1999b). The genetic mechanism is unknown, but it threatens the stability of selected cultivated strains and suggests, again, that multispore spawns may be too genetically diverse to be a reliable means of distributing specific strains. Thus, strain degeneration is a serious problem faced by the *V. volvacea* industry. Although the fungus is homothallic, isolates collected from widely separated localities were found to be genetically different (Chiu *et al.*, 1995). Strain degeneration in *L. edodes*, however, is a different natural phenonemon. It arises from the contamination of airborne spores and random mating events between spore germlings (monokaryons) and the resident dikaryon. The resultant somatic incompatibility and the changed genomes account for the decrease in crop yield in later flushes.

In contrast to the edible fungi described above, *Ganoderma* is unique as a mushroom which is cultivated for its medicinal value. Global production of *Ganoderma* in 1997 was about 4300 t (about 3000 t of which were grown in China (Moore and Chiu, 2001). Under the names *lingzhi* (in Chinese) or *reishi* (in Japanese), several *Ganoderma* spp. of the *Ganoderma lucidum* complex provide various commercial health drinks, powders, tablets, capsules and diet supplements. *Ganoderma* is highly regarded as a traditional herbal medicine, and its popularity in China has spread to other Asian countries, and also to the wider world. *Ganoderma* is cultivated by being inoculated into short segments of wooden logs which are then covered in soil in an enclosure (often a plastic-covered tunnel) which can be kept moist and warm. The fruit bodies then emerge in large numbers quite close together and the conditions encourage the fungus to form the desirable long-stemmed basidiome. Like *Volvariella*, *Ganoderma* also expresses considerable developmental and morphological plasticity, but in this case the fruit bodies are very polymorphic in the wild. The outcome is taxonomic confusion; over 70 species have been described, many being invalid because they are simply developmental or morphological variants of the type species. Collected fruit bodies are highly prized, but if the mycologist cannot identify them reliably, the confusion can be used to profit from exaggerated claims or fraudulent products. Detailed analysis, including DNA markers, will provide reliable means of identification of both fruit bodies and, more importantly, the processed products. The safety of *Ganoderma* products, many of which are sold for prophylactic use and long-term consumption, is another significant aspect being researched, including the ability of fungal fruit bodies to accumulate heavy metal ions. *Lingzhi*, unlike other commonly cultivated mushrooms, is a root rot pathogen and collections can be made in urban and rural environments, including roadsides. Thus, although the wild-collected *lingzhi* is popularly thought to have better medicinal properties, bioaccumulation of heavy metals from polluted environments by this fungus has been demonstrated and will definitely pose a hazard. Artificial cultivation instead of collection from the wild would be a better strategy. Public health concerns would then be similar to those affecting other cultivated crops and would be satisfied by random sampling and quality assessment to assure the safety of the food product in the market.

In traditional Chinese medicine, *Ganoderma* is known as 'the mushroom of immortality', which is claimed to alleviate or cure virtually all diseases. Current research is focused on purification and characterization of the bioactive components and determination of clinical value, especially putative anti-tumour and anti-ageing properties. Tests with mice showed *Ganoderma* extract to be only a modest dietary supplement. Evidence of genotoxic chromosomal breakage or cytotoxic effects by *Ganoderma* extract has not been found, nor protection against the toxic effects of the radiomimetic mutagen ethyl methanesulfonate. The true medical value of *Ganoderma* extract needs further investigation, though it is essentially safe for consumption (Chiu *et al.*, 2000).

Acknowledgements

The authors thank the UGC for earmarked grants supporting some of the projects dealing with life cycle strategy in *V. volvacea* and diversity of Chinese *L. edodes*. S.W.C. thanks the Chinese University of Hong Kong for providing direct grants which partially supported the research work on use of spent mushroom substrate in degrading pollutants, and carbon metabolism of *Pleurotus*. We are grateful to Miss May Cheng and Miss Annie Law for their contribution to the projects on spent mushroom substrate and heavy metal uptake by *Pleurotus*.

References

Al-Gharawi, A. and Moore, D. (1977) Factors affecting the amount and the activity of the glutamate dehydrogenases of *Coprinus cinereus*. *Biochimica et Biophysica Acta* 496, 95–102.

Anonymous (1997) *The Chinese Forests*. Forestry Department, Beijing.

Balis, D., Chatjipavlidas, J. and Flouri, F. (1991) An integrated system of treating liquid wastes from olive oil mills. In: Balis, C. (ed.) *Proceedings of the Symposium on Treatment of Olive Oil Mills Wastes*. Geotechnical Chamber of Greece and International Council of Olive Oil, Chania, Greece, pp. 66–74.

Bobek, P., Ozdin, L. and Kuniak, L. (1994) Mechanism of hypocholesterolemic effect of oyster mushroom (*Pleurotus ostreatus*) in rats – reduction of cholesterol absorption and increase of plasma cholesterol removal. *Zeitschrift für Ernährungswissenschaft* 33, 44–50.

Burnett, J.H. (1968) *Fundamentals of Mycology*. Edward Arnold, London.

Chang, S.T. (1993) Mushroom biology: the impact on mushroom production and mushroom products. In: Chang, S.T., Buswell, J.A. and Chiu, S.W. (eds) *Mushroom Biology and Mushroom Products*. The Chinese University Press, Hong Kong, pp. 3–20.

Chang, S.T. (1998) A global strategy for mushroom cultivation – a challenge of a 'non-green revolution'. In: Lu, M., Gao, K., Si, H.-F. and Chen, M.-J. (eds) *Proceedings of the '98 Nanjing International Symposium, Science and Cultivation of Mushrooms.* JSTC-ISMS, Nanjing, pp. 5–15.

Chang, S.T. and Miles, P.G. (1991) Recent trends in world production of cultivated mushrooms. *Mushroom Journal* 504, 15–18.

Chihara, G. (1993) Medical aspects of lentinan isolated from *Lentinus edodes* (Berk.) Sing. In: Chang, S.T., Buswell, J.A. and Chiu, S.W. (eds) *Mushroom Biology and Mushroom Products.* The Chinese University Press, Hong Kong, pp. 261–266.

Chiu, S.W. and Moore, D. (1999a) Carbon metabolism in the Oyster mushroom, *Pleurotus pulmonarius.* In: Sietsma, J.H. (ed.) *Abstracts, Seventh International Fungal Biology Conference.* University of Groningen, The Netherlands, p. 46.

Chiu, S.W. and Moore, D. (1999b) Segregation of genotypically diverse progeny from self-fertilized haploids of the Chinese straw mushroom, *Volvariella volvacea. Mycological Research* 103, 1335–1345.

Chiu, S.W., Moore, D. and Chang, S.T. (1989) Basidiome polymorphism in *Volvariella bombycina. Mycological Research* 92, 69–77.

Chiu, S.W., Chen, M.-J. and Chang, S.T. (1995) Differentiating homothallic *Volvariella* mushrooms by RFLPs and AP-PCR. *Mycological Research* 99, 333–336.

Chiu, S.W., Ma, A.M., Lin, F.C. and Moore, D. (1996) Genetic homogeneity of cultivated strains of shiitake (*Lentinula edodes*) used in China as revealed by the polymerase chain reaction. *Mycological Research* 100, 1393–1399.

Chiu, S.W., Ching, M.L., Fong, K.L. and Moore, D. (1998a) Spent Oyster mushroom substrate performs better than many mushroom mycelia in removing the biocide pentachlorophenol. *Mycological Research* 102, 1553–1562.

Chiu, S.W., Chan, Y.H., Law, S.C., Cheung, K.T. and Moore, D. (1998b) Cadmium and manganese in contrast to calcium reduce yield and nutritional value of the edible mushroom *Pleurotus pulmonarius. Mycological Research* 102, 449–457.

Chiu, S.W., Yip, M.L., Leung, T.M., Wang, Z.W., Lin, F.C. and Moore, D. (1998c) Genetic diversity in a natural population of shiitake (*Lentinula edodes*) in China. In: *Abstracts of the Sixth International Mycological Congress, Jerusalem, Israel, 23–28 August.* IMC, Jerusalem.

Chiu, S.W., Wang, Z.M., Chiu, W.T., Lin, F.C. and Moore, D. (1999a) An integrated study of individualism in *Lentinula edodes* in nature and its implication for cultivation strategy. *Mycological Research* 103, 651–660.

Chiu, S.W., Yip, P. and Moore, D. (1999b) Genetic diversity of *Lentinula edodes* collected from Hubei Province, China. In: Sietsma, J.H. (ed.) *Abstracts, Seventh International Fungal Biology Conference.* University of Groningen, The Netherlands, p. 47.

Chiu, S.W., Wang, Z.M., Leung, T.M. and Moore, D. (2000) Nutritional value of *Ganoderma* extract and assessment of its genotoxicity and antigenotoxicity using comet assays of mouse lymphocytes. *Food and Chemical Toxicology* 38, 173–178.

Chorvathova, V., Bobek, P., Ginter, E. and Klvanova, J. (1993) Effect of the oyster fungus on glycemia and cholesterolemia in rats with insulin-dependent diabetes. *Physiological Research* 42, 175–179.

Corner, E.J.H. (1932a) A *Fomes* with two systems of hyphae. *Transactions of the British Mycological Society* 17, 51–81.

Corner, E.J.H. (1932b) The fruit-body of *Polystictus xanthopus* Fr. *Annals of Botany* 46, 71–111.

Corner, E.J.H. (1966) A monograph of the cantharelloid fungi. *Annals of Botany Memoirs* 2, 1–255.

Elhiti, M.M.Y., Butler, R.D. and Moore, D. (1979) Cytochemical localization of glutamate dehydrogenase during carpophore development in *Coprinus cinereus*. *New Phytologist*, 82, 153–157.

Ewaze, J.O., Moore, D. and Stewart, G.R. (1978) Co-ordinate regulation of enzymes involved in ornithine metabolism and its relation to sporophore morphogenesis in *Coprinus cinereus*. *Journal of General Microbiology* 107, 343–357.

Fermor, T.R. and Wood, D.A. (1981) Degradation of bacteria by *Agaricus bisporus* and other fungi. *Journal of General Microbiology* 126, 377–387.

Flegg, P.B. (1996) The shape and form of mushrooms. *Mushroom Journal* 560, 12–13.

Flouri, F., Soterchos, D., Ioannidou, S. and Balis, C. (1995) Decolourization of olive oil liquid wastes by chemical and biological means. In: *Proceedings of the International Symposium on Olive Oil Processes and By-Products Recycling, Granada, Spain.*

Hadar, Y., Kerem, Z. and Gorodecki, B. (1993) Biodegradation of lignocellulosic agricultural wastes by *Pleurotus ostreatus*. *Journal of Biotechnology* 30, 133–139.

Hammad, F., Ji, J., Watling, R. and Moore, D. (1993a) Cell population dynamics in *Coprinus cinereus*: co-ordination of cell inflation throughout the maturing fruit body. *Mycological Research* 97, 269–274.

Hammad, F., Watling, R. and Moore, D. (1993b) Cell population dynamics in *Coprinus cinereus*: narrow and inflated hyphae in the basidiome stipe. *Mycological Research* 97, 275–282.

Kalisz, H.M., Moore, D. and Wood, D.A. (1986) Protein utilisation by basidiomycete fungi. *Transactions of the British Mycological Society* 86, 519–525.

Kalisz, H.M., Wood, D.A. and Moore, D. (1989) Some characteristics of extracellular proteinases from *Coprinus cinereus*. *Mycological Research* 92, 278–285.

Kerem, Z., Friesem, D. and Hadar, Y. (1992) Lignocellulose degradation during solid state fermentation – *Pleurotus ostreatus* versus *Phanerochaete chrysosporium*. *Applied and Environmental Microbiology* 58, 1121–1127.

Loh, J., Randers, J., MacGillivray, A., Kapos, V., Jenkins, M., Groombridge, B., Cox, N. and Warren, B. (1999) *The Living Planet Report*. WWF International, Gland, Switzerland.

Mackinnon, J., Sha, M., Cheung, C., Carey, G., Zhu, X. and Melville, D. (1996) *A Biodiversity Review of China*. WWF International, Hong Kong.

Martinez, A.T., Gonzalez, A.E., Valmaseda, M., Dale, B.E., Lambregts, M.J. and Haw, J.F. (1991) Solid-state NMR studies of lignin and plant polysaccharides degradation by fungi. *Holzforschung* 45, 49–54.

Moore, D. (1981) Evidence that the NADP-linked glutamate dehydrogenase of *Coprinus cinereus* is regulated by acetyl-CoA and ammonium levels. *Biochimica et Biophysica Acta* 661, 247–254.

Moore, D. (1998) *Fungal Morphogenesis*. Cambridge University Press, New York.

Moore, D. and Chiu, S.W. (2001) Fungal products as food. In: Pointing, S.B. and Hyde, K.D. (eds) *Bio-Exploitation of Filamentous Fungi*. Fungi Diversity Press, Hong Kong, pp. 237–265.

Moore, D. and Ewaze, J.O. (1976) Activities of some enzymes involved in the metabolism of carbohydrate during sporophore development in *Coprinus cinereus*. *Journal of General Microbiology* 97, 313–322.

Moyson, E. and Verachtert, H. (1991) Growth of higher fungi on wheat straw and their impact on the digestibility of the substrate. *Applied Microbiology and Biotechnology* 36, 421–424.

Novak Frazer, L. (1996) Control of growth and patterning in the fungal fruiting structure. A case for the involvement of hormones. In: Chiu, S.W. and Moore, D. (eds) *Patterns in Fungal Development*. Cambridge University Press, Cambridge, pp. 156–181.

Ortega, G.M., Martinez, E.O., Betancourt, D., Gonzalez, A.E. and Otero, M.A. (1992) Bioconversion of sugarcane crop residues with white-rot fungi *Pleurotus* spp. *World Journal of Microbiology and Biotechnology* 8, 402–405.

Pegler, D.N. (1996) Hyphal analysis of basidiomata. *Mycological Research* 100, 129–142.

Redhead, S.A. (1987) The Xerulaceae (Basidiomycetes), a family with sarcodimitic tissues. *Canadian Journal of Botany* 65, 1551–1562.

Saiz-Jimenez, C. and Gomez-Alarcon, G. (1986) Effects of vegetation water on fungal microflora. *Actas International Symposium on Olive By-Products Valorization.* Food and Agricultural Organization of the United Nations (FAO), United Nations Development Programme, Sevilla, Spain, pp. 61–76.

Stewart, G.R. and Moore, D. (1974) The activities of glutamate dehydrogenases during mycelial growth and sporophore development in *Coprinus lagopus* (*sensu* Lewis). *Journal of General Microbiology* 83, 73–81.

Tan, Y.-H. and Moore, D. (1992) Convenient and effective methods for *in vitro* cultivation of mycelium and fruiting bodies of *Lentinus edodes*. *Mycological Research* 96, 1077–1084.

Tan, Y.-H. and Moore, D. (1994) High concentrations of mannitol in the shiitake mushroom (*Lentinula edodes*) *Microbios* 79, 31–35.

Tan, Y.-H. and Moore, D. (1995) Glucose catabolic pathways in *Lentinula edodes* determined with radiorespirometry and enzymic analysis. *Mycological Research* 99, 859–866.

Tomati, U., Galli, E., Dilena, G. and Buffone, R. (1991) Induction of laccase in *Pleurotus ostreatus* mycelium grown in olive oil waste waters. *Agrochimica* 35, 275–279.

Tripathi, J.P. and Yadar, J.S. (1992) Optimisation of solid substrate fermentation of wheat straw into animal feed by *Pleurotus ostreatus* – a pilot effort. *Animal Feed Science and Technology* 37, 59–72.

Umar, M.H. and van Griensven, L.J.L.D. (1995) Morphology of *Agaricus bisporus* in health and disease. In: Elliott, T.J. (ed.) *Science and Cultivation of Edible Fungi.* A.A. Balkema, Rotterdam/Brookfield, The Netherlands, pp. 563–570.

Umar, M.H. and van Griensven, L.J.L.D. (1997a) Morphogenetic cell death in developing primordia of *Agaricus bisporus*. *Mycologia* 89, 274–277.

Umar, M.H. and van Griensven, L.J.L.D. (1997b) Hyphal regeneration and histogenesis in *Agaricus bisporus*. *Mycological Research* 101, 1025–1032.

Umar, M.H. and van Griensven, L.J. L.D. (1997c) Morphological studies on the life span, developmental stages, senescence and death of fruit bodies of *Agaricus bisporus*. *Mycological Research* 101, 1409–1422.

Umar, M.H. and van Griensven, L.J.L.D. (1998) The role of morphogenetic cell death in the histogenesis of the mycelial cord of *Agaricus bisporus* and in the development of macrofungi. *Mycological Research* 102, 719–735.

Umar, M.H. and van Griensven, L.J.L.D. (1999) Studies on the morphogenesis of *Agaricus bisporus*: the dilemma of normal versus abnormal fruit body development. *Mycological Research* 103, 1235–1244.

Upadhyay, R.C. and Hofrichter, M. (1993) Effect of phenol on the mycelial growth and fructification in some basidiomycetous fungi. *Journal of Basic Microbiology* 33, 343–347.

Van Loon, P.C.C., Sonnenberg, A.S.M., Swinkels, H.A.T.I., van Griensven, L.J.L.D. and van der Heijden, G.W.A.M. (1995) Objective measurement of developmental stage of white button mushrooms (*Agaricus bisporus*). In: Elliott, T.J. (ed.) *Science and Cultivation of Edible Fungi*. Balkema, Rotterdam, pp. 703–708.

Zhang, J., Wang, G.Y., Li, H., Zhuang, C., Mizuno, T., Ito, H., Suzuki, C., Okamoto, H. and Li, J.X. (1994) Antitumor polysaccharides from a Chinese mushroom, Yuhuangmo, the fruiting body of *Pleurotus citrinopileatus*. *Bioscience, Biotechnology and Biochemistry* 58, 1195–1201.

Index

Page numbers in **bold** refer to illustrations and tables